ROUTLEDGE LIBRARY EDITIONS: WW2

Volume 23

PARTISAN WARFARE

PARTISAN WARFARE

OTTO HEILBRUNN

LONDON AND NEW YORK

First published in 1962 by George Allen & Unwin Ltd

This edition first published in 2022
by Routledge
2 Park Square, Milton Park, Abingdon, Oxon OX14 4RN

and by Routledge
605 Third Avenue, New York, NY 10158

Routledge is an imprint of the Taylor & Francis Group, an informa business

© 1962 George Allen & Unwin Ltd

All rights reserved. No part of this book may be reprinted or reproduced or utilised in any form or by any electronic, mechanical, or other means, now known or hereafter invented, including photocopying and recording, or in any information storage or retrieval system, without permission in writing from the publishers.

Trademark notice: Product or corporate names may be trademarks or registered trademarks, and are used only for identification and explanation without intent to infringe.

British Library Cataloguing in Publication Data
A catalogue record for this book is available from the British Library

ISBN: 978-1-03-201217-9 (Set)
ISBN: 978-1-00-319367-8 (Set) (ebk)
ISBN: 978-1-03-210495-9 (Volume 23) (hbk)
ISBN: 978-1-03-210501-7 (Volume 23) (pbk)
ISBN: 978-1-00-321557-8 (Volume 23) (ebk)

DOI: 10.4324/9781003215578

Publisher's Note
The publisher has gone to great lengths to ensure the quality of this reprint but points out that some imperfections in the original copies may be apparent.

Disclaimer
The publisher has made every effort to trace copyright holders and would welcome correspondence from those they have been unable to trace.

PARTISAN WARFARE

Otto Heilbrunn

with a Foreword by
Colonel the Hon. C. M. Woodhouse
D.S.O., O.B.E., M.P.
*Commander of the British Military
Mission to Greece*, 1943-4

LONDON
GEORGE ALLEN & UNWIN LTD
RUSKIN HOUSE MUSEUM STREET

FIRST PUBLISHED IN 1962

This book is copyright under the Berne Convention. Apart from any fair dealing for the purpose of private study, research, criticism or review, as permitted under the Copyright Act, 1956, no portion may be reproduced by any process without written permission. Enquiry should be made to the publisher.

© *George Allen and Unwin Ltd,* 1962

PRINTED IN GREAT BRITAIN
in 11 *on* 12 *pt. Ehrhardt*
BY SIMSON SHAND LTD
LONDON, HERTFORD AND HARLOW

AUTHOR'S NOTE

A good deal of the source material discussed in this book was made available to me at the Imperial War Museum Library in London. I wish to record my sincere thanks to the Librarian, Miss R. E. B. Coombs, for her help, and to the Library staff for meeting my heavy demands on the Museum's bookstore with unfailing cheerfulness, beyond the call of duty.

FOREWORD BY C. M. WOODHOUSE

Dr Heilbrunn has already established himself as a historian of irregular warfare. But the subject is not merely a matter of past history, because the so-called 'nuclear stalemate', which has made total warfare improbable, has at the same time made limited warfare the only kind that the world can afford to risk. One hopes, naturally, that the risk will be avoided; but since even a conventional war of the traditional, pre-nuclear kind might easily lead unintentionally up to a total war between great powers and is therefore also likely to be avoided, there remains the residual danger of what may be called 'sub-conventional' warfare in marginal areas, which the great powers would be free to support or disown, to fan up or suppress, according to their immediate interpretation of their own interests. Such are the outbreaks which we have seen in recent years in Malaya, Vietnam, Algeria, Cyprus, Cuba, Laos and elsewhere. These are also, if Korea proves, as we hope, to have been the last conventional war between major powers, the kinds of war we must expect to see renewed in the future.

The Resistance during the second world war was the prelude to this new kind of warfare. It was not, of course, a new invention between 1940 and 1945: one remembers, on the contrary, the Spanish resistance during the Napoleonic Wars, which gave us the word *guerrilla* to add to our language, and the exploits of Lawrence and others during the Arab Revolt of 1917. But these were side-shows (Lawrence's own word) in support of a major conventional war, without which they would have achieved practically nothing. Since the second world war, the corresponding outbreaks of irregular warfare have stood on their own as the major, if not the only, armed conflicts in their particular struggle, not a side-show in support of a major war elsewhere. The Spanish Civil War of 1936–8 is their archetype. Irregular warfare has accordingly become more professional and highly organized. It has had to acquire a sense of strategy, not merely of tactics. Perhaps eventually it will drop the epithet 'irreg-

FOREWORD

ular'. Even by 1945 the 'partisans' of southern Europe and the Balkans had ceased to so describe themselves, and adopted instead the nomenclature of regular armies.

Those who fought with the partisans of the second world war will find that already there have been profound changes in the evolution of partisan warfare since 1945. But thanks to Dr Heilbrunn's keen sense of the continuity of that evolution, they will also recognize their own side-shows as forming an integral part of the history of this fascinating subject. He does us the honour of frequent quotation from our accounts of war-time experience; and it is encouraging to find that the lessons of that experience have been confirmed by later application elsewhere. His book is perhaps the first comprehensive study of the theoretical aspects of partisan warfare, at least in the English language. It is firmly grounded in practice, and likely to serve for a long time as a standard work.

CONTENTS

FOREWORD	page 9
INTRODUCTION	13

Chapter

1 THE BIRTH OF A PARTISAN MOVEMENT — 15

The Structure of Partisan Movements in China, Soviet Russia and Yugoslavia—The Inception of a Partisan Movement: its aims, structure, appointments, recruiting, intelligence, supplies, funds, theatre of war, relations with rival movements, propaganda, arms, and training.

2 THE CHANCES OF SURVIVAL — 32

Anti-Guerilla Intelligence—Morale and Support of the Population—Economic Warfare.

3 THE CHANCES OF SUCCESS: STRATEGIC ASPECTS — 40

Strategic Role of Base Areas—The Need for Semi-Regular Forces.

4 THE CHANCES OF SUCCESS: OPERATIONAL ASPECTS — 53

Guerilla Intentions: Aggressiveness versus 'Pillbox Psychology'; the Wars in China, Indo-China and Soviet Russia—Anti-Guerilla Intentions: Isolation of the Enemy (Malaya); Annihilation through *Jagdkommandos* (Soviet Russia), Pseudo-Gangs (Kenya) and Encirclement (Yugoslavia)—Partisan Intelligence.

5 GUERILLA AND ANTI-GUERILLA TACTICS — 78

Guerilla Tactics: The Viet-minh Manual—Anti-Guerilla Tactics: Defence—Surprise Attack and Hunt—Armoured Trains and Tanks.

6 GUERILLA AND ANTI-GUERILLA TECHNIQUES — 107

Guerilla Techniques for Sabotage, Ambushes and Attacks—Anti-Guerilla Technique for Encirclement.

CONTENTS

7 RELATIONS WITH THE REGULAR ARMY   111

The various Forms—The Case for Advance Preparation—Subordination or Co-ordination?—Table of Organization—Officers or Partisans as Leaders?—Partisans and Special Forces.

8 THE AIR FORCE IN GUERILLA WARFARE   123

In World War II: Guerilla Air Support by the Royal Air Force, the American Air Force and the Red Air Force. Anti-Guerilla Air Support by the Luftwaffe—Post-War developments: Anti-Guerilla Air Support in Greece, Malaya, Indo-China, Cyprus and Algeria—The Significance of Air Support for Ground Operations—Joint Planning and Control—Airbase Security.

9 GUERILLAS AND NUCLEAR WARFARE   140

Guerilla Objectives—Guerilla Intelligence and Sabotage—Guerilla Immunity from Nuclear Attack.

10 THE TREATMENT OF GUERILLAS AND THE POPULATION   143

The Treatment of Guerillas: The Law. Prisoner of War Status?—The Treatment of the Population: The Law. Counter-Terror? Collective Punishment. Winning the Support of the Population.

11 ON WHOSE SIDE VICTORY?   159

The Doctrine of 'la Guerre révolutionnaire' – The Theory and the Praxis—Prolonged or Shortened War—A Summing-up.

12 APPENDIX: ON PARTISAN WARFARE IN WORLD WAR II   171

BIBLIOGRAPHY   187

INDEX   194

MAPS

Greece: Partisan Areas in 1947	50
China: Partisan Areas in 1946	54
Northern Indo-China	61
Yugoslavia: Partisan Areas in 1944	72

INTRODUCTION

In the last war guerillas were never decisively beaten by regular armies. On the contrary, guerillas usually managed to inflict considerable losses upon the opposing forces. The guerilla successes appear all the more impressive if one considers that Germany, while defeating in quick succession the regular armies of a number of countries, did not succeed in annihilating their guerilla forces: Norway, Denmark, France, Yugoslavia, Albania and Greece are obvious examples. During the first part of the German-Russian campaign the Red Army suffered many defeats, yet during the same period the Soviet guerillas had more successes than reverses.

Hence the modern guerilla was almost invested with the nimbus of invincibility, that is until Great Britain beat guerillas in Malaya, Kenya and Cyprus.

It is, of course, true that most of the guerilla forces operating in the last war against Germany became truly effective only after her resources had been stretched to the utmost and her fighting strength weakened, but this does not apply to her campaign against Russia prior to Stalingrad. Did Great Britain, then, discover the secret of anti-guerilla warfare, a secret which has apparently eluded the French in Indo-China and Algeria? When is the military machine superior to guerilla opposition and under what conditions does the advantage lie with the guerillas? What is required to turn the tables?

No doubt, the British success in Malaya was to some extent due to the skilful use of propaganda and bribes, penetration played its part in Kenya, while the helicopter was the war-winning weapon in Cyprus. Can general rules be devised for the successful conduct of anti-guerilla warfare, or are there specific tactics applicable to each theatre of war? If so, can one at least establish general rules for the treatment of guerillas?

And closely allied to this question, can general rules be devised for the treatment of the population? Does severity pay, does leniency not create its own dangers, what is meant by just treatment of the

INTRODUCTION

population, and can justice be reconciled with one's own security? Can anti-guerillas shorten the duration of prolonged guerilla wars?

These are some of the questions which we try to answer in this book. It deals with guerilla and anti-guerilla warfare and takes its examples from recent campaigns. It tries to indicate some general rules of partisan and anti-partisan war and therefore differs from other guerilla or anti-guerilla books, which usually describe a particular campaign or the author's share in it. The present work is written in the belief that guerilla movements play a part in modern wars regardless of whether they are national or revolutionary, nuclear or conventional, global or localized.

CHAPTER I

THE BIRTH OF A PARTISAN MOVEMENT

During World War II partisan movements fought in many theatres of war. Apart from the Chinese, none of them had been prepared beforehand; they all came into being during the war. That they proved useful to their own side is not in doubt. It seems surprising therefore that they were not organized before the outbreak and that their inception was frequently left to chance, particularly as this dilatoriness caused many additional problems, one of them being that of leadership.

The supreme partisan leaders were not as a rule commissioned by the State authorities and the leaders who emerged had either to seek accommodation with local chieftains, much to the movement's detriment, as Mihailovitch had to in Yugoslavia—he was confirmed only afterwards, by his government in exile—or they tried to eliminate competing leaders and movements—in Albania, Yugoslavia and Greece—or, together with other partisan movements in the country, they voluntarily submitted to higher authority, which could in the nature of things exercise only lose control, as in France. Only exceptionally was it possible for the State to appoint a supreme leader in the course of the war and effectively subordinate regional and local partisan leaders. This was, of course, the case in Soviet Russia. Time was on her side.

Most probably the internecine struggle in Albania, Yugoslavia, Greece and the Philippines would not have been avoided even if attempts at forming a unified command had been made there before the war. However, as a rule, it seems dangerous to defer the formation of a loyal partisan movement until hostilities have broken out, and some of the dangers become obvious when we study the cumbersome process of its inception by reviewing some practical examples from the recent past. This is one purpose of this chapter. It has another aim, of equal military relevancy. Since the best chance of destroying a partisan movement is to suppress it at the outset, the opponent must be familiar with the technique of its inception, so that he can diagnose the signs and act in time; we try here to demonstrate this technique.

If we disregard adventurers and brigands, discontent and patriotism are the two mainsprings of partisan movements. The discontented, the revolutionary guerillas, fight against the established order—represented by a 'colonial' or 'imperialist' government of foreigners or a 'feudalist' or 'bourgeois' or bolshevist government of their own nationals—while the patriots fight against the foreign invader of their country. Yet the dividing line between the two is not always clearly marked: Mao Tse-tung's guerillas were revolutionaries by origin and fought a revolutionary and national war at the same time during the Japanese invasion from 1937 on, and so did Tito in Yugoslavia during World War II. And again, during World War II, some Russian guerillas fought against the Germans in order to win independence for their region, and others, with the same aim, attacked the Soviets. But although the categories do not always fit and counter-guerillas have no place in the scheme at all, most guerilla movements conform to type and the distinction has a certain usefulness.

The first of the large modern revolutionary guerilla movements sprang up in China, and internal and external conditions were favourable to its inception and early growth. The Chinese Communist Party was inaugurated in 1921. In the following year the first official visitor from Moscow arrived in the person of Adolf Joffe. He had previously been Soviet Ambassador to Germany, whence he was expelled for carrying out revolutionary propaganda; guerilla war does not seem to have been his line; the Chinese Communist Party certainly did nothing then to organize an army. This task was left to the next Soviet mission, which was led by Borodin and included military advisers. The official purpose of this mission was to assist Sun Yat-sen in drawing up a plan for democratic reform in China. During his four-year stay in China, until his expulsion in 1927, Borodin rendered three vital services to the Chinese Communists: he succeeded in getting the Chinese Communist Party recognized and affiliated with the Kuomintang; he established in China agrarian Communist areas which formed the manpower reservoir for the partisans; and he arranged for the training of Chinese Red officers at the Military

BIRTH OF A PARTISAN MOVEMENT

Academy. It was, as Mao said in 1938, the communist 'participation in the work of the Whampoa Military Academy' which made the Communist Party begin 'to see the importance of military affairs'. Soon afterwards the communists, in Mao's words, 'got hold of some armed forces': an independent regiment under a communist general seems to have been the first of such acquisitions, followed soon by a division officered mostly by communists. Mao himself had recruited his first guerilla 'division' of peasants, miners and students, and on this basis the Chinese Red Army was formed in 1927. Its organization, practicable only in so vast a country, is of little interest here.

The outstanding patriotic guerilla movement of World War II was, of course, that of Soviet Russia. Nowhere were conditions for its inception more favourable. There was in Russia a lawful government, resident in the country and with all the might of State and party behind it; there was an unbeaten Army, a large reservoir of manpower in the occupied areas from which partisans could be raised; much of occupied Russia was excellent guerilla territory—the forests, the swamps and the mountains; and from the vast unoccupied area the partisans could be reinforced and supplied. But that was not all. Soviet Russia had frequently been warned of the impending German attack—by Great Britain,[1] the United States,[2] the Sorge Ring in Japan,[3] the 'Red Orchestra' in Switzerland,[4] and the Resistance in Czechoslovakia,[5] and had time to organize the guerillas beforehand. In the party and NKVD echelons in every locality it had ready-made recruiting offices; even after the invasion these echelons remained intact in all areas in the path of the German advance and party cells were still functioning in German occupied country, although reduced in efficiency by their underground existence and mass arrests of their functionaries. Finally, the Soviets had had fair experience of guerilla fighting in the days of the Civil War. Yet, in spite of all these advantages, they found it difficult to organize their partisans. Practically nothing had been done before the out-

[1] Sir Winston Churchill, *The Second World War*, vol. iii, London 1950, p. 320, and vol. iv, London 1951, p. 443.
[2] *The Memoirs of Cordell Hull*, vol. ii, London 1948, p. 968.
[3] Major-General C. A. Willoughby, *Sorge: Soviet Master Spy*, London 1952, p. 84.
[4] W. L. Flicke, *Agenten funken nach Moskau*, Kreuzlingen 1954, pp. 86, 89, 95.
[5] Professor Vladimir Krajina, 'La Résistance tchécoslovaque,' in *Cahiers d'Histoire de la Guerre*, February 1951, No. 3; also D. J. Dallin, *Soviet Espionage*, New Haven 1955 p. 134.

break and it took them eighteen months of war to build up an efficient movement.

Immediately after Stalin had made his famous appeal to the people on July 3, 1941, for the formation of partisan units 'in areas occupied by the enemy', the first attempt at organizing the partisan movement was made by the Main Administration of Political Propaganda of the Red Army.[1] The choice of this agency seems curious and its policy was certainly unorthodox. By using every available channel—the commissars with the Army, the Central Committee of the Communist Party and the NKVD—it tried to create the greatest possible number of detachments, without thought for centralization or co-ordination, over-all organization or chain of command. At operational level the most glaring shortcoming was the partisan commanders' lack of training. It was not yet realized that partisan intelligence could be a vital source of information for the Red Army; partisan missions were thought of purely in terms of ambushes and sabotage.

Two months later the partisan movement was reorganized and the Central Staff of the Partisan Movement was formed under Marshal Voroshilov. It was independent of Army and NKVD and under the control of the Central Committee of the Communist Party but naturally army personnel became more numerous on the Staff and Red Army officers were sent to the partisans to take command of bands.[2] Some partisan leaders were taken to unoccupied Russia and trained, wireless operators and equipment were dropped to the more important detachments, a partisan intelligence network was set up and the first tenuous liaison was established between some detachments and the Red Army.[3] But it was not until 1943 that the movement found its final shape. Voroshilov had in the meanwhile been replaced by a party man, Ponomarenko, and the partisans were now organized almost like a regular army. Since they were fighting on the other side of the front but in ever closer association with the Red Army, the chain of command went down via the Central Staff, through partisan staffs with the higher commands of the Red Army,

---

[1] Major Edgar M. Howell, *The Soviet Partisan Movement 1941–1944*, Department of the Army Pamphlet, Washington 1956, p. 47.

[2] Major E. M. Howell, *loc. cit.*, pp. 65, 80 seq.

[3] Major E. M. Howell, *ibid.*, Brigadier C. A. Dixon and O. Heilbrunn, *Communist Guerilla Warfare*, London, New York, 1954, pp. 48 seq., 73, 77 and 81, and O. Heilbrunn, *The Soviet Secret Services*, London, New York, 1956, pp. 49 seq.

and then across the front line through partisan commands in the enemy's rear. Army intelligence officers were attached to each partisan echelon[1] and by now practically all detachments had received field radio equipment.

Because the Red Army withstood the German onslaught, the Soviets were given the chance to experiment with their partisan movement until they could draw effective support from it for their regular army in the field. Of all the countries invaded during the last war only China had a similar chance.

While circumstances favoured the partisan movement in Russia, they seem to have conspired against Colonel Mihailovitch's guerillas in Yugoslavia. He was confronted by a Quisling government at home and supported—for a while—by a government in exile without power or ability to help. His rightful claim to leadership was frequently ignored by his friends and denied by his foe. He had no unoccupied zone to draw on, and, unlike Tito, no underground movement for propaganda, recruitment and supplies; the Chetniks on whom he depended for recruits were divided in their loyalty. Finally, while the brunt of the German attack in Russia was borne not by the partisans but the Red Army, all German operations in Yugoslavia were launched against Mihailovitch's guerillas and Tito's partisans; they were the only Yugoslav fighting forces in the field.

During the fight of the Yugoslav Army against the Germans, Mihailovitch was Chief of Staff of a motorized division in Bosnia. He refused to capitulate and, when he was routed by the Germans, tried to break through with a number of officers and men to join other fighting remnants of the Army. He escaped capture and moved on with a small group to Serbia and the wooded heights of Ravna Gora, where other units also sought refuge. They fought on as guerillas under Mihailovitch. He built up his movement by direct recruiting among the peasants and Chetniks,[2] by sending his officers to take over command of local peasant resistance groups which had sprung up all over the country, by establishing liaison with other officers who had formed their own guerilla detachments, and by bringing under his command—nominally at least—such local chieftains and their bands as would agree to his leadership.

Tito's movement was formed on different lines. After the military

---

[1] Major E. M. Howell, *loc. cit.*, pp. 137 seq.
[2] An organization of World War I veterans.

commission of the Central Committee of the Yugoslav Communist Party, presided over by its Secretary-General Tito, had decided to prepare organizationally and politically for the fight, military committees were formed in the provinces, regions, districts and localities under the direction of the party. Their first tasks were to find the arms stores left by the Yugoslav Army after its capitulation and to start recruiting among the party members, members of the Communist Youth Union, and certain non-communist patriots. Immediately after the invasion of Russia the military commission, which had in the meanwhile adopted the new title of 'military committee', was renamed General Staff of the Partisans, with Tito in command.[1]

The Soviet Russian partisan organization differed basically from both Yugoslavian organizations because in Russia there was a regular army in the field and in Yugoslavia there was none. Hence in Russia liaison between army and partisans had to be established and their actions co-ordinated, partisan intelligence had to be utilized for the army and, since the partisans were never required for or supposed to engage in orthodox fighting, there was no need to build up the formations above brigade level.

Yet Mihailovitch's and Tito's organizations differed in important ways. Both soon organized their forces into brigades and divisions, and Tito later formed corps and armies. But Mihailovitch's forces were all assigned fixed geographical areas, while Tito had his Proletarian Brigades at all times, and others frequently, available for employment anywhere in the country. Apart from the military committees at all levels, Tito also had intermediate commanders in the field while Mihailovitch had none. As a result, Mihailovitch's forces, unlike Tito's, could not count, when hard pressed, on reinforcements from GHQ or anywhere else, or on diversionary attacks by other units. For recruiting and logistics Mihailovitch's local commands had little support from above; they had to rely on their own arrangements in practically every respect, and in their turn frequently failed to keep GHQ informed, becoming subordinate in name only. Why all this was so is outside our scope; the cohesion of the movement was loose, there was little unity or discipline. The organizational set-up not only reflected these defects, but also sharply accentuated them. The ulti-

---

[1] For the above cf. Lt.-Col. Brajus Kovic-Dimitrye, 'La Guerre de Libération Nationale en Yougoslavie (1941-45),' in *European Resistance Movements 1939-1945*, Oxford etc., 1960, pp. 302 seq.

BIRTH OF A PARTISAN MOVEMENT

mate result was that Mihailovitch's movement broke largely because of its structural weaknesses, while Tito's organization withstood all vicissitudes and grew stronger and stronger.

It is therefore evident how important the structure is to the efficient functioning of the movement and how carefully the partisans must make their choice. Conversely, it is no less important for the anti-partisans to ascertain the structure of the opposing guerilla movement at the earliest possible moment, in order to seek out its leaders, destroy the organization, and, while it is still in its formative stages, to recognize its aims.

How, then, is a partisan movement created? If there is no leader appointed by the government, or if the movement is not based on an army turned guerilla, there is no standard procedure. Hardly any two movements go through the same processes and not many turn out alike in shape. Still, while choice and circumstances differ, all movements conform in a varying degree to the following scheme.[1]

Like any other joint venture, a guerilla movement starts with a meeting of a few people with similar outlook, army officers, former guerilla or civil war fighters, students, workers or peasants, intellectuals or party leaders, patriots or traitors, who want to 'do something about it', whether it be the foreign invasion or the real or imagined oppression of the country or the natives or the common people. An unorganized general uprising is rare, and the lone wolf who takes on himself the whole burden of forming a movement is undesirable. His early arrest will destroy the organization—in Rome, and elsewhere too, the first partisan movement became badly unbalanced for this reason—and while one of a group may be, or soon become, the driving force, there ought always to be, right from the start, a second-in-command or trusted deputies to take over in case of accident. The members of the group should be influential or inspiring, secretive and circumspect. They need not necessarily be officers, but it helps at this stage if one at least is a regular soldier and another experienced in clandestine work.

The organizers all have the same aim—fighting the foreign invader

[1] During the last war the Special Operations Executive in Britain sent more than 500 agents to enemy occupied countries either to form sabotage groups or to make contact with existing resistance movements. SOE had 60 training schools in the UK, its means of transport, material, and many radio communication reception stations. In Singapore 101 Special Training School had a decisive effect on the formation of the war-time resistance in Malaya.

or fighting the armed forces of the government, or both; but they must now decide on ways and means. In fighting the foreign invader they may specialize in one type of guerilla war, to the exclusion of all others: the war-time forces in Denmark limited themselves to sabotage and intelligence, and others elsewhere had much more restricted aims. In Turin, for example, action committees staged armed insurrections in factories, electricity works, etc., with the sole object of driving the Germans out of their particular works;[1] in Holland the 'Aussen Ministerium' was a student organization for the clandestine repatriation of Dutch slave-labour students in Germany, and another group, the Nationaal Steun Fund, was set up in order to raise funds for the Resistance.[2] Sometimes such organizations widened their activities later: in Italy the movement first formed for the protection of Rome from German destruction soon supplied intelligence to the Allies[3] and finally developed into the military organization of the entire Italian resistance.

While restricted-action groups are at this stage faced with relatively few problems, organizers deciding on full-scale partisan war may be confronted with a difficult choice. Tito chose to harass the Italians and Germans from start to finish, without waiting for the tide to turn. So did Mihailovitch before him, but the terrible retribution meted out by the Germans against his countrymen made him decide to limit the activities of his force until an Allied army was approaching or entering Yugoslavia. This change of mind was fatal to his movement and coupled with its structural weakness, led by stages to its disintegration.[4] No guerilla movement in the field can afford to remain inactive for long; by so doing it loses its morale and sense of purpose. When patriots meet to discuss the formation of a guerilla force they must realize that the enemy will react with every means at his disposal. Only if they and the country can take it, should they proceed with the immediate formation of the movement. Otherwise it will do more harm to their country by its existence than to the

---

[1] Cf. G. Vaccarino, 'Le Mouvement de Libération Nationale en Italie (1943-45),' in *Cahiers d'Histoire de la Guerre*, February 1951, No. 3, p. 102.

[2] P. D. Vollgraff, 'La Résistance en Hollande,' in *Cahiers d'Histoire de la Guerre*, loc. cit., p. 44.

[3] Cf. Ronald Seth, *The Undaunted, The Story of Resistance in Western Europe*, London 1956, p. 302.

[4] Cf. O. Heilbrunn, *The Soviet Secret Services*, pp. 104-122, and *Der sowjetische Geheimdienst*, Frankfurt am Main, 1956, pp. 100-118.

enemy by its effectiveness. Communist organizers, however, hardly have to face this problem. As Mr Julian Amery has pointed out, they may welcome reprisals because these add to the number of the dispossessed, their potential supporters.[1]

If the future movement is intended to fight the armed forces of the government, the conspirators must assess their chances of victory in the field. They must for this purpose answer a number of questions: is the government force loyal, what is its morale, can they hope one day to match their opponent, what is the attitude of the population, and so on. The Chinese communists and the Viet-minh believed in their superiority and were right; the leaders of the Greek post-war rebellion and the Malayan Races Liberation Army came to the same conclusion and were wrong. If the leaders cannot count on defeating the opposing forces in the field, can they at least apply pressure, rouse world public opinion, turn it against the government and make the government leave the country in consequence? This novel technique was tried out in turn with varying results by the terrorists in Palestine, Tunisia, Kenya, Cyprus,[2] and probably also in Algeria. None of their leaders could ever have thought of defeating the opposing British or French troops, but all hoped to last long enough in the field to achieve their aims by other means.

Next, the organizers must decide on the structure of the movement. Not all of them do. When the first Soviet partisan detachments made for the mountains and forests nobody had yet worked out a table of organization, and the forest gangs in Kenya were never subject to higher direction or control. Normally, however, their structure will depend in the first place on whether they are fighting in support of their army in the field. In that case they must arrange for liaison with the army or subordination under the army. However, if they are fighting independently because there is no army for them to support, or because the army opposes them, they may choose to penetrate an existing organization, as the Mau Mau leaders did by using the Kenya African Union.[3] This process requires much time and must be started long before the movement begins its operations. As a rule the leaders must build up their own organization. Communist

[1] Julian Amery, *Sons of the Eagle, A Study in Guerilla War*, London 1948, p. 168.

[2] This interpretation is given in regard to Cyprus by Dudley Barker in *Grivas, Portrait of a Terrorist*, London 1959.

[3] For an authoritative study see Colonial Office, *Historical Survey of the Origins and Growth of Mau Mau*, by F. D. Corfield, Her Majesty's Stationery Office, London 1960.

revolutionary movements as well as nationalist ones—such as the Algerian—sometimes start by forming propaganda cadres; they then set up local organizations, and apply terror methods, and only afterwards are guerilla forces recruited.

If the movement is supposed to be purely military, as were the non-communist guerilla formations in Albania, Yugoslavia, Greece and elsewhere, the task of selecting a suitable structure is, in theory at least, not a complex one. There must be a chain of command, and the number of links depends on the number and size of the detachments and their geographical spread. But if the movement is at the same time political and military, there are three possibilities: either the politicians are in command, or the political and military sides function separately, or the military are in charge. Let us illustrate these different possibilities.

Case No. 1: the politicians are in control. This is usual in communist movements, but there are two variants: there may be one or two chains of command. In post-war Malaya there was one chain of command, going down from the Politbureau to the Central Executive Committee which exercised control over the North, Central and South Federation. Under each Federation followed, in descending order, the State Committees, the District and Branch Committees, and finally the Cells. The fighting units were organized by regiments, companies, platoons and sections, the regiments being attached to the State Committees and the subordinate units to the subordinate Committees and Cells.[1] But if, under the aegis of the Communist Party, liberated areas are set up in revolutionary wars, there may be one chain of command for the military and another chain for the civil organization, and this scheme was adopted during the post-war Greek rebellion; the stationary civil administration took care of political matters and supplies in the bases while the bands under separate command were free to move from sector to sector.

If the military commander happens to be the party boss, as Tito was, his powers are for all practical purposes unrestricted. But if he is not a communist, and the ELAS commander in war-time Greece was not, special safety devices are adopted; the High Command there consisted of a triumvirate, the military commander and two communists, and in all major decisions the three had to act together. Party control ensured that in this war on two fronts, the national and the

---

[1] Cf. H. Miller, *Menace in Malaya*, London 1954, pp. 103 f.

BIRTH OF A PARTISAN MOVEMENT

revolutionary, the revolutionary struggle was vigorously conducted.

Case No. 2, the military and political side exist independently of each other: the Italian war-time resistance developed in this direction. A number of local Roman committees formed before and after the armistice joined together in a Committee of National Liberation, comprising party representatives from communists to Christian Democrats and set up a Military Committee. In June 1944 the CNL succeeded in combining the various partisan groups into one organization, the Corps of Volunteers for Liberation (CVL), the Military Committee in Milan, consisting of delegates from various movements, was transformed into the General Command of the Corps, and a general was appointed Chief of the Corps. Unified commands, consisting of movement delegates, were also established at regional and local levels. Military resistance thus lay in the hands of the CVL. At the same time a Central Committee of Liberation was set up; it was responsible for civilian resistance and local administration before the arrival of the Allies.[1] Although the communists tried to exercise their influence in the North, the CVL remained in fact outside party politics.

Finally Case No. 3, in which the military element takes charge, is exemplified by the Irgun (National Military Oranization) which operated in Palestine after the last war as a terrorist organization. Its five-man High Command, advised by a General Staff, took command of the 'divisions'—a few dozen regulars and many part-timers—and controlled group political activities as well.[2]

We therefore find that there are the following seven basic tables of organization of a guerilla movement [see pages 26–29 below].

If the organizers have not already done so, they must now appoint the commander. Next, they have to decide who will be responsible for recruiting, commissions, intelligence, supplies, funds and the allocation of theatres of operation, whether it shall be the organizers themselves, the guerilla command, a special or the party organization.

Recruiting is usually done by all available means, that is by the guerillas themselves and, if the movement is connected with any other agency, by that agency as well. In Russia partisans were recruited in the occupied areas not only by the bands but also by the so-called feeder bands and the underground party organization,

[1] Cf. Ronald Seth, *loc. cit.*, p. 309, and G. Vaccarino, *loc. cit.*, pp. 89 f.
[2] Menachem Begin, *The Revolt*, London 1951 (?), pp. 3 and 61.

PARTISAN WARFARE

A. PARTISANS FIGHTING WITH THE REGULAR ARMY

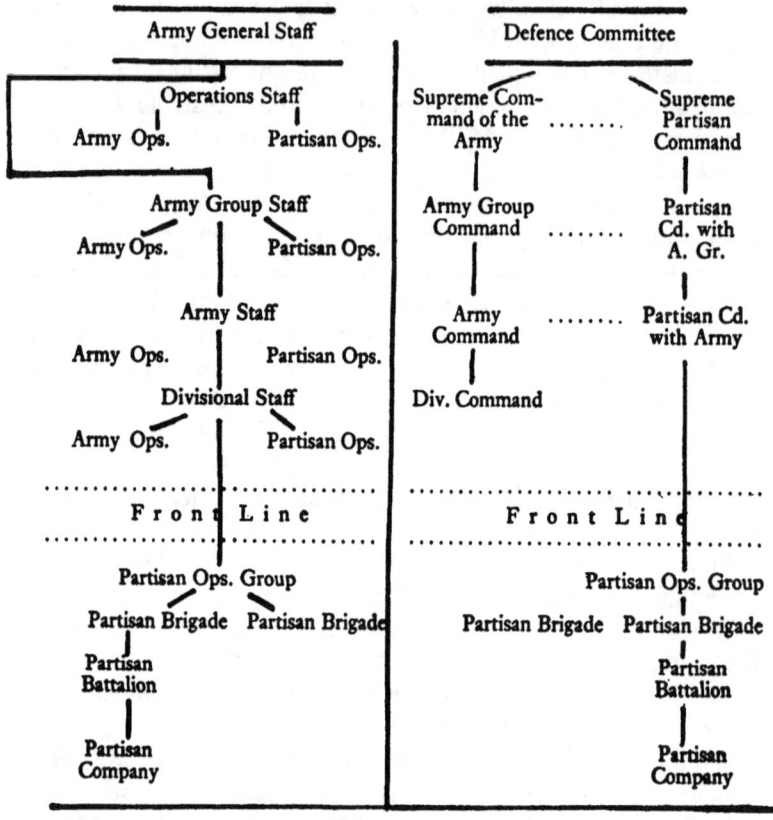

CHART 1: Partisan Movement subordinated to the Army

CHART 2: Partisan Movement coordinated with the Army

while Partisan GHQ and the NKVD sent units from unoccupied Russia. Tito's partisans were recruited by the local Military Committees of the party, in war-time Greece ELAS and EAM emissaries went to towns and villages, and in Malaya the original recruits sent in 1941 to 101 Special Training School in Singapore had all been picked by the party, while later recruiting was done by the party Military Affairs Committee in each State. The Viet-minh sent party agents to the villages outside French control and used the party organization for

BIRTH OF A PARTISAN MOVEMENT

B. PARTISANS FIGHTING INDEPENDENTLY OF THE REGULAR ARMY

1. The Movement has only Military Aims

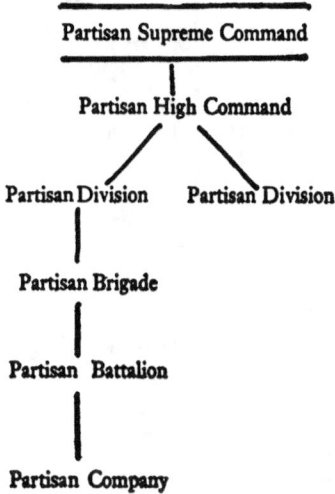

CHART 3

recruiting in places held by the French, and in Kenya the Land Freedom Armies were supported by the Mau-Mau district and local committees. Here too the partisans did their own recruiting everywhere as well. Sometimes parts of other organizations were taken over in Russia dispersed Army units and the destruction battalions were absorbed into the partisan movement; Tito received considerable support from former Mihailovitch followers; in Italy Carabinieri, armed forces which had evaded internment by the Germans, and British and other Italian war prisoners who had escaped during the armistice joined the guerillas; and ELAS even obtained its commander from a rival guerilla movement. Recruits, it should be noted, were not always volunteers, and frequently a considerable part of the force consisted of part-timers who worked in their normal civilian occupation when not called up. The communists did not restrict themselves to recruiting only communists; they drew the line somewhere—'reactionaries' were excluded—but otherwise accepted any

2. The Movement has Political and Military Aims

a. The Politicians are in Command

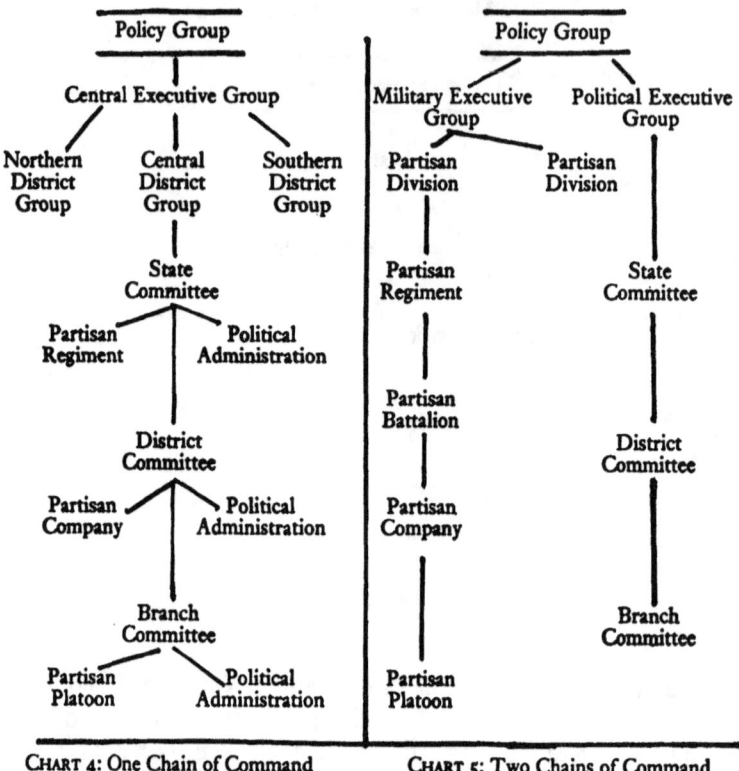

CHART 4: One Chain of Command CHART 5: Two Chains of Command

support as long as they could keep control of the movement and indoctrinate its members.

The organizers should also decide whether a special supply organization should be created. In much of occupied Europe great quantities of supplies came to the partisans through the British Special Operations Executive and later also the American Office of Strategic Services, and the question of a special supply organization in the country itself seldom arose. In Russia, however, GHQ, the party centres and the feeder bands already mentioned helped to some extent; ELAS was also supplied during the war through ETA and during the post-war rebellion by its local civil administration; and in post-

BIRTH OF A PARTISAN MOVEMENT

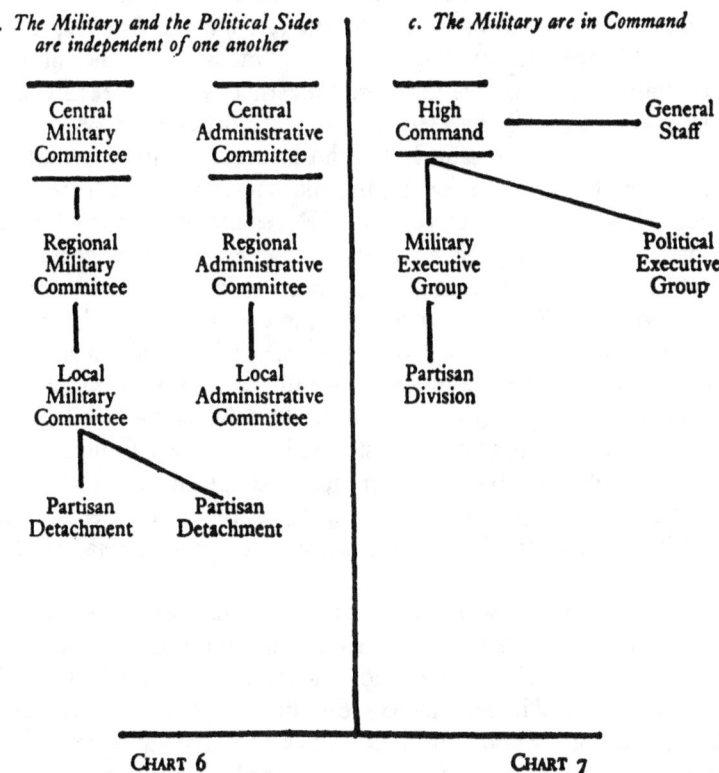

Chart 6   Chart 7

war Malaya the Min Yuen was responsible for supplying the Malayan Races Liberation Army. Generally speaking, a local supply organization seems useful where the enemy is denying food to the guerillas or where they are fighting far away from inhabited or cultivated areas.

Special arrangements by the organizers are also called for if the guerillas are to receive their supplies from abroad. Much of the remaining items of the agenda, apart from raising funds and possibly the appointment of Staff officers and detachment leaders, is left to the guerilla commander, the General Staff or the military committee, but there are still two points in particular to which the organizers must give attention: the propaganda line and relations with rival movements.

We shall deal with propaganda later and discuss here only relations with rival movements. Often the decision will have to be taken at a later stage when operations have started and clashes occur. The choice is wide as was seen during the last war. In Albania, Yugoslavia, Greece and the Philippines they fought each other and the communists were frequently more concerned with suppressing their rivals than fighting the Germans; in Poland the Home Army was eliminated by other means. In Czechoslovakia and Greece temporary armistices between the communist and non-communist forces were secured through the intervention of Beneš and the British Military Mission respectively. Sometimes area agreements were concluded. In Greece unified command was established for one particular operation, while the non-communist forces in Czechoslovakia decided on a proper division of labour, one movement specializing in sabotage, the second in collecting intelligence and transmitting it to London, and the third engaging in propaganda. In Albania the Royalists first co-operated with the Communists and then with the Centre, and in Belgium only the extreme Left groups united, while in France, Italy, Holland, and temporarily in Greece, all groups put themselves under a unified command; the communists, however, always retained their separate identity.

Needless to say, where there are such unresolved rivalries the enemy will try to exploit them, as indeed he must. In Albania the Germans gave arms to one movement so that it could fight the communists, and in Yugoslavia the Italians negotiated with Chetnik units about arms and supply deliveries and actually seem to have sent a battalion to fight alongside Mihailovitch's forces against Tito's,[1] while the Germans agreed with Chetnik units to keep the big Danube bend free from Tito's partisans.[2] Oddly enough, when the Germans had the chance of ending the war in Yugoslavia altogether by backing Mihailovitch against Tito, they are said to have refused to take it.[3]

---

[1] Cf. *The Trial of Mihailovic*, Stenographic Records and Documents, Belgrade 1946, pp. 197 and 447.

[2] Cf. Major Herbert Kriegsheim, *Getarnt, Getäuscht und doch Getreu, Die geheimnisvollen 'Brandenburger'*, Berlin 1958, Part II, written by an anonymous German senior officer, p. 312.

[3] *Ibid*, p. 313, where it is related that between Mihailovitch and the commander of the 4th Brandenburg Regiment an agreement was concluded to the effect that Mihailovitch was to subordinate himself to the Germans if he were given command of an antibolshevist Division of Serbs on the eastern front. He is also reported to have offered the services of several hundred thousand fighters against Tito. The agreement was to last until American troops landed on the Adriatic coast, since he could then no longer

BIRTH OF A PARTISAN MOVEMENT

But to revert to the birth of a partisan movement, the guerillas usually themselves assume control of intelligence for their own purposes, although in post-war Malaya the Min Yuen was charged with this task. If the guerillas fight alongside their regular army, army officers may be seconded to the guerillas for collecting and transmitting army intelligence; this was the arrangement ultimately adopted in Soviet Russia.

Finally, revolutionary guerilla movements must also lay the groundwork for the civil administration and the establishment of base areas; we shall discuss the general pattern in Chapter 3 below.

The guerillas must now prepare camps and storage places in their area, they must obtain arms—all movements have somehow managed to obtain at least a few—move off to their location and receive their training. Communications with HQ, other units and their own army must be arranged. Then action can start.[1] It may be modest at first, designed to create confidence: mine laying, sabotage, attacks on unguarded or poorly guarded targets, on small enemy patrols, on small posts and convoys. Minor enemy forces in the area are mopped up, but engagements are as yet avoided.

The guerilla movement is born.

---

guarantee the reliability of his forces. When the German High Command was notified of the agreement it ordered it to be annulled. This incident was not mentioned in Mihailovitch's trial. See also Paul Leverkuehn, *Der geheime Nachrichtendienst der deutschen Wehrmacht im Kriege*, Frankfurt am Main 1957, pp. 124 f.

[1] The importance of having some regular soldiers to form the backbone of a detachment is stressed in the review of the Strategic Problems of the Malayan Revolutionary War by the Politbureau of the Malayan Communist Party (1949). The Malayan Communist Party complains there that, unlike the revolutionary forces of other nations, it is handicapped by the lack of revolutionary members of the regular army. The review is reprinted in Gene Z. Hanrahan, *The Communist Struggle in Malaya*, New York 1954, p. 119.

CHAPTER 2

THE CHANCES OF SURVIVAL

In the last war a partisan movement's chances of survival were high and had it not been for the struggle between communists and non-communists, hardly any guerilla movement would have become a casualty at all. Yet is seems doubtful whether this would still be true in a nuclear conflict. We have seen how long it took to rectify the defects of some of the movements created after the outbreak of the last war: there were faulty chains of command and weak structures, rivalries between movements and a lack of trained band-leaders, intelligence was neglected at first and communications were poor. But there was ample time to correct and improve. If a nuclear Blitzkrieg should happen no partisan movement formed after the outbreak of hostilities would have the opportunity of making any worthwhile contribution before the end and it would probably never take the field at all.

Apart from this, anti-guerilla intelligence has learned its lesson and the chances of detecting a partisan movement, during its creation and and in its formative stages, have improved. While the Germans did not even consider the possibility of ever meeting organized partisan resistance anywhere, least of all in Russia, we now know that as long as their allies keep on fighting, few countries will submit to the humiliation of defeat, and that communists outside Russia will take up the partisan fight if she is involved in war. Equally, the appearance of colonial terrorists is no longer an unexpected phenomenon. Anti-guerilla intelligence is now forewarned and better prepared for its task.

Anti-guerilla intelligence in enemy country and in colonial territories is conducted on the same lines. In either case there must be agencies, widely spread, for collecting information and these agencies are the military intelligence or, in colonies, military intelligence, the

Special Branch and District Commissioners or their equivalent. Both types depend on the same sources of information—their own observations, their contacts with the population in general and 'V-men' (*Vertrauensleute* or confidential agents) in particular, and interrogations. Ground intelligence may be supplemented by air intelligence, but never supplanted; the French forces had to find this out in Indo-China. In partisan areas anti-partisan commandos are indispensable for reconnaisance, as is indicated by German experience in World War II.

During the last war the Germans set up Information Collecting Centres wherever partisans fought against them, and the German *Manual for Warfare against Bands* laid great stress on the need for the immediate evaluation, by people of experience and knowledge of local conditions, of all incoming reports on partisans.[1] Mr F. D. Corfield emphasizes the same point in the Colonial Office *Survey of the Origins and Growth of Mau Mau* which we have already quoted; the persons assessing the reports must be able to appreciate their short- and long-term implications.[2] Mr Corfield goes on to emphasize the need for a systematic and periodical assessment of the reports—a function performed in war-time Germany by the Army General Staff, Foreign Armies East and West and by Himmler's organization —and the importance of proper control and direction of intelligence, including the need to transmit intelligence downwards as well as upwards.

'The result of the evaluation has to be followed at once by decisions', says No. 56 of the German Manual. In colonial territories where the formation of a partisan movement is only suspected, the decision will naturally be left to higher authority. This is not the case if partisan warfare has already started in enemy or colonial territory; as the Manual appropriately states (in No. 56),

'If the evaluation shows that immediate action is required and possible, the order for action has to be given and subsequently the report has to be passed on to the immediate superior together with a report on the action taken.

'If the evaluation shows that action is required but the necessary forces are not available, the report has to be passed on to the immedi-

---
[1] Issued by the High Command of the Armed Forces, May 6, 1944, No. 55.
[2] Pp. 31 f.

ate superior with the recommendation that action is considered necessary, but that our own units are not available.'

Intelligence must never be misled into thinking that it can rely on precedents: in Greece, Malaya and Indo-China the post-war guerillas all adopted a structure different from that of their war-time movements, and when in Kenya the members of the Mau Mau Central Committee were arrested, their successors on the Council of Freedom set up quite a different organization.

The Colonial Office *Survey* convincingly shows that a guerilla movement can be detected even before its members begin action. Unlike the military in enemy territory, the colonial administration has of course the advantage of being on its home grounds and with established contacts. But if this is the criterion, then, as far as wartime guerillas are concerned, their chances of successfully forming a movement gradually decrease the longer a vigilant enemy has been in occupation.

If the partisans can avoid detection and annihilation while their movement is being formed, their expectation of life is rather good. This was shown not only by practically all war-time, and also the successful post-war, partisan movements, but by the defeated ones as well; the emergency in post-war Greece lasted three years, in Malaya twelve years, and in Kenya and Cyprus four years.

The reason for this apparently remarkable endurance is not far to seek: owing to the peculiarities of this kind of war it takes a long time for the anti-guerillas to turn the tables. The anti-guerillas' superiority in heavy arms and equipment frequently does not count at all, because it cannot be brought to bear on the partisans in their chosen theatre of operations—jungle, marshes, mountains, forests. Anti-guerilla wars are largely won by other means; morale and communications. Colonel C. M. Woodhouse first elucidated their significance and interdependence as follows:

'Morale and communications ... are ambivalent, both cut both ways. There has never been a successful guerilla war conducted in an area where the populace is hostile to the guerillas, and conversely it is virtually impossible to stamp out a guerilla war in an area where the populace continues to support the guerillas. The art of defeating the guerillas is therefore the art of turning the populace against them (as the Greeks successfully demonstrated in 1949). And here rises the second important factor of communications. In an area of primitive communications the anti-guerilla force is hopelessly handicapped unless it can get the populace on its side; and it will only get the populace on its side if it can protect them against the guerillas—which it can only do if it has good communications. Conversely, the guerillas can only survive in the long run if they have the necessary communications to ensure supplies of arms and ammunition—whether from abroad by sea and air, as in the second world war, or across a friendly land frontier, as in Greece in 1947–49.'[1]

It takes time to turn the populace against the guerillas. Hence their comparative longevity.

In the battle for support the guerillas usually start with an advantage: they are natives and know the population; they can exploit real or imaginary grievances, and they can always count on the loyalty of a sizeable part of the population, the Malayan aborigines, the Kikuyu, the Greek Cypriots. If elements of the population are not willing they are coerced: in Greece families were taken as hostages and children abducted, Mau Mau took forcible oaths and started with an attack on loyal Africans, in Cyprus the first victims were Greeks. The pattern is always the same: indoctrinate or subjugate the population. It is too familiar to require further description.

The guerillas have another advantage: they can get in first with their propaganda and practically have the field to themselves until they begin military operations. In the indoctrination of the population the guerilla forces themselves play a vital part, following the example set long ago by the Eighth Route Army in China,[2] and they in turn are subject to 'ceaseless political instruction with the object of improving both individual and collective morale and of securing a proper attitude towards the population'.[3]

[1] In *The Twentieth Century*, June 1954, p. 571; cf. also by the same author 'The Greek Resistance 1942–44,' in *European Resistance Movements 1939–1945, op. cit.*, p. 388.
Equally, the first task of the British during the Cyprus troubles was to improve communications.
[2] Cf. Mao Tse-tung, 'Interview with the British Correspondent James Bertram,' reprinted in *Selected Works of Mao Tse-tung*, vol. ii, London 1954, p. 96.
[3] Tito in a conversation with Brigadier Sir Fitzroy Maclean, reprinted in his *Disputed Barricade*, London 1957, p. 235. Général L.-M. Chassin, 'Guerre en Indochine,' in *Revue de Défense Nationale*, xvii, 1953, p. 9, reports that the Viet-minh forces received propaganda training for eight hours daily for several months, with lectures on the slavery

The anti-guerillas cannot always hope to win the battle for the support of the population. In that case they try to neutralize it. Nor can they always hope to defeat the guerillas by military means alone. In that case they try to corrupt them. The anti-guerillas' tactics are therefore these:

A. In regard to the population

1. It must be protected against the guerillas, preferably by way of self-help through home guards and militia, and supported by anti-guerilla forces which can come quickly to their assistance.

2. It must be converted, if possible, preferably with the help of loyal members of the population. Mao Tse-tung spread his propaganda 'by word of mouth, by leaflets and bulletins, by newspapers, books and pamphlets, through theatrical performances and the films, through schools, through mass organizations and through cadres',[1] and now radio and propaganda recordings played from aircraft have been added. In Malaya, where the majority of the aborigines at first shielded the communists, the formation of an aborigine fighting regiment by New Zealanders must have contributed much towards winning the confidence of this part of the population. Yet the best propaganda will fall on deaf ears unless it is backed by military successes against the guerillas.

3. Not all can be converted and the process takes a long time. The hostile population must therefore be isolated. French forts in Vietnam proved ineffective for the control of the population because there were too few of them, their garrisons were too small and they clung too much to the forts. The German Manual on *Warfare against Bands* had already stressed that the fight against partisans must be aggressive; this being the case in Malaya, the jungle forts built there by the government fulfilled their purpose. So did the resettlement of the Chinese squatters from the jungle fringe where they had supplied the Liberation Army. Temporary evacuation may also be useful. In Cyprus houses were sealed for months and cleared of the inhabitants if no witnesses came forward to testify about murders in the neighbourhood, and house curfews, internment and the taking of hostages

of the West, colonialism and exploitation, the history of imperialism and the misery of the working classes in the West, the triumph of justice and truth through communism, the rudiments of communist doctrine and the history of bolshevist revolution, Mao's fight against Chiang, and of the Viet minh against France.

[1] 'On the Protracted War,' in *Selected Works, op. cit.*, p. 205.

are other means of control. The danger with many of these isolation measures is that they will alienate the population and make the task of winning it over more difficult, if not impossible. They must therefore be used with discretion. We shall discuss this aspect later, in Chapter 10.

B. In regard to the guerillas

1. Propaganda can be successful here only if morale was never high or has already been weakened, through successes by the anti-guerillas, lack of support by the foreign allies of the partisans, or desertion by the population. The successes of the anti-guerillas need not all be military; it is also demoralizing for the guerillas if their opponent denies them arms and food, and the food denial measures taken in Malaya were outstandingly successful. In Kenya, however, before the declaration of the emergency, the Kikuyu had too easy access to arms[1] and in Cyprus the rebels found it possible to get arms through the mails, even after the establishment of postal censorship, by penetrating the service.[2]

2. An incentive was provided in Malaya in the form of rewards to partisans who surrendered, and a number of them went back in order to talk other guerillas into surrendering also. This system 'finally flushed the jungle of communists',[3] because they realized that they had lost the war.

In the last resort, counter-propaganda and other measures help to shorten the war if, and only if, the anti-guerillas are successful on the battlefield.

Another factor which can considerably extend the life span of the guerilla movements is the economic one. If the theatre of war is a country which is underdeveloped or not well off and if it is not assisted by richer allies, it can ill afford the financial expenditure of an anti-guerilla war and may therefore have to restrict its war effort.

It is indeed one of the avowed aims of revolutionary guerillas to extend the war to the economic front and the means of economic warfare are manifold. In Kenya and Cyprus a boycott of British goods was organized; in Malaya and Kenya the insurgents en-

---

[1] *Survey of the Origins and Growth of Mau Mau, op. cit.*, p. 225 f.
[2] Cf. Dudley Barker, *op. cit.*, p. 157.
[3] London *Times*, 'Emergency in Malaya declared at an End.' August 1, 1960.

deavoured to secure the withdrawal of labour from plantations, mines and farms, and loyal workers were intimidated; in Greece the partisans tried to create food shortages by raids on villages and the removal of livestock; and generally private property is destroyed, public utilities, especially communications, are sabotaged and public life is further disrupted by the murder of administration officials. Last but not least, defence expenditure is driven up.[1]

The guerillas expect a double return from crippling the country's economy: the impact of counter-measures, such as taxation increases, reduction of unemployment benefits, security measures etc., should increase the number of their sympathizers, and the government might be unable militarily or politically to keep up the fight.

That this type of warfare has its own dangers for the guerillas was shown in Malaya when the Politbureau of the Malayan Communist Party changed its strategy in 1951 and laid down that in order 'to win the masses the party must... stop burning new villages,... attacking reservoirs, power stations and other public services.... Rubber trees, tin mines and factories must not be destroyed because of the resentment of the workers who lose their employment'.[2] But on the whole, economic warfare can inflict considerable damage on the guerillas' opponent and had it not been for the financial help given first by Great Britain and then the United States to the post-war Greek Government it may be doubted whether it could have seen the war through. Even there, however, the guerillas extended the duration of the war by their attack on the country's economy.

At any rate, in this type of warfare, the guerillas have all the advantages as long as they avoid alienating potential supporters: it is easier and requires less manpower to strike at the objectives than to protect them, and the guerillas are invulnerable in this type of warfare; they do not have an economy.

We thus find that the guerillas' chances of survival are rather good, because their opponent is not necessarily superior to them on the battlefield and he is at a disadvantage in the conduct of psychological and economic warfare, which plays a bigger part in guerilla wars than in conventional warfare.

[1] On the French side the two war years 1952 and 1953 in Indo-China cost no less than $4,680 million. Cf. Colonel J. F. McQuillen, 'Indochina,' *Marine Corps Gazette* 1955, p. 56.

[2] Reprinted in Gene Z. Hanrahan, *op. cit.*, p. 130, and London *Times* of December 1, 1952.

THE CHANCES OF SURVIVAL

We shall now survey the guerillas' chances of success, and we deal with the strategic aspects in Chapter 3 and the operational and tactical aspects in Chapters 4 and 5. Not all guerilla movements have strategic aims.

In fact, only independent partisan movements have strategic aims, while auxiliary partisan movements have only tactical aims, as we will now try to show.

1. Independent Partisan Movements. They are the only fighting forces confronting the enemy in the theatre of war, either because the regular army of the country never entered the war or has ceased to fight, or because the partisans are fighting against it. The guerillas fighting in post-war Malaya against England or in North Africa against France were independent forces. Independent partisans may first apply guerilla methods, but sooner or later they must fight like a regular force in order to achieve their aims, and their strategy must then be that of a regular force. They are fighting at the front. They are the front. Their aim is to defeat the enemy in the field. They therefore have strategic, in addition to merely tactical, aims.

2. Auxiliary partisan movements. They fight in association with the regular forces—though on opposite sides of the front—against the enemy. The Soviet guerillas of the last war, fighting in support of the Red Army, were auxiliaries. Auxiliary partisan movements do not have the task of defeating the enemy in the field. They do not fight battles, they avoid engagements, they do not try to throw back the enemy or to hold territory. They do not fight at, but behind the front. Their function is to weaken the enemy; to defeat him is the regulars' task. Auxiliary partisans therefore have no strategic aims; their tasks are operational or tactical.[1]

Hence the next chapter, which discusses strategic aspects, is only concerned with independent partisan movements, while the inquiry into the operational and tactical aspects in the subsequent chapters deals with independent as well as auxiliary movements.

[1] Compare for the above O. Heilbrunn, *Partisanenbuch*, Zurich 1960, pp. 12 f.

CHAPTER 3

THE CHANCES OF SUCCESS: STRATEGIC ASPECTS

Mao Tse-tung was the first to treat guerilla battlecraft as a proper subject of military science and nobody has made a greater contribution to the understanding of guerilla strategy than he.

Probably few revolutionary partisan movements have had such an easy birth and none so hard a life as the Chinese. We have briefly referred to its origins in Chapter 1.[1] From the beginning almost to the end it was inferior, and most of the time vastly inferior in numbers or *matériel* or both, to its Chinese Nationalist and Japanese enemies, and yet it succeeded in sustaining itself over so long a period until final victory almost a quarter of a century later.

How was it done? After what is known as the Long March—Chiang Kai-shek calls it appropriately the long flight—the communists had been greatly weakened in membership and fighting strength and it is open to doubt whether the continuation of their policies—armed uprisings, sovietization and forceful dispossession of landlords—would have retrieved their fortunes. Then came the Japanese invasion. Not only did it give relief from Chinese Nationalist attacks but, in taking up the fight against the Japanese,[2] the communists could harness the patriotism of those in regions under their control and draw on the support of the many Chinese who had suffered grievously at Japanese hands. They could also incorporate into their ranks nationalist army stragglers and small formations which had been cut off from their units. They only had to give their policies a temporary new look, which they proceeded to do. One of the principles of the so-called political work of their forces was the establishment

---

[1] For Mao's campaigns see Général L.-M. Chassin, *L'Ascension de Mao Tse-Tung* (1921-1945), Paris 1953, and *La Conquête de la Chine par Mao Tse-Tung* (1945-1949), Paris 1952.

[2] For the Communist reasons see Elliot R. Goodman, *The Soviet Design for a World State*, New York 1960, p. 84.

of unity between the army and the people, and this was now interpreted to mean that the people's property must be respected; the crops had to be safeguarded against the enemy, especially during harvesting. Commerce was to be protected. The forces were only allowed to requisition a limited amount of provisions and supplies and had to rely considerably on their own efforts on the land and in small factories. Finally they had to carry out propaganda among the masses and to organize and arm them. Instead of being more feared than the existing authorities, they had now set out to be more respected.

How successful the transformation of the communist party image was, soon became apparent in the revival of the party fortunes: in 1937, sixteen years after the formation of the Communist Party, its membership had shrunk to 40,000 organized members and 30,000 partisans; by 1940, after two or three years of communist fighting against the Japanese, both figures had increased almost twentyfold. In 1944 party membership was about 1,000,000 and about two and a half million men were under arms.

This build-up of the forces was only possible because the communists recognized the importance of guerilla bases and knew how to make use of them. From the time of Borodin they had always been in possession of large areas where they openly exercised political power and therefore, strangely enough for a revolutionary movement, the conspiratorial element was almost entirely absent. Now, in the enemy's rear and in suitable locations, they devoted their energies to transforming zones fought over by both sides, the so-called guerilla districts, into guerilla bases, i.e. nucleus States within the State. The first and fundamental requirement for success was the availability of a detachment strong enough to annihilate or decisively defeat the enemy. The second was the defeat of the enemy. When local victory was achieved, the population, with guerilla help, was roused to mass anti-Japanese struggles. This was the third condition for the establishment of a base: the awakening of the masses. 'Through these [anti-Japanese] struggles', Mao has explained, 'the people are to be armed and home guards and additional guerilla detachments formed. Through these struggles mass organizations of labourers, peasants, young people, women, children, merchants and professionals are to be formed and developed, keeping step with the degree of the people's political consciousness and their heightening urge to struggle. The power of the people cannot be manifested without organization.'[1] The puppet administration was deposed and replaced by an anti-Japanese (communist) administration which controlled the area, passed laws, collected taxes,[2] and raised troops.

Once the base was established it had to be strengthened and developed. At certain times, as Mao explained, emphasis was laid on development, which meant the extension of the guerilla district and the enlargement of the guerilla detachment, at other times the emphasis was on strengthening, i.e. awakening and organizing the masses, including training of guerilla and local armed forces.

'Guerilla warfare cannot survive for long and develop itself without a base', says Mao. But for the facilities for guerilla bases 'provided by nature for us to exploit' and the enemy's insufficiency in armed forces, the communist guerillas would have been condemned to a 'roaming and raiding' war; 'but no war of this type ever succeeded'. As it was, they drew from these bases their power of survival and growth.

It must be recognized that bases of this type are only possible where the country is vast or regions are inaccessible; elsewhere the partisans have to make do with hide-outs and camps. Malaya was obviously a country where the communist Chinese could emulate Mao's example and endeavour to establish liberated areas.

Indeed, the first efforts of the Malayan communists in this direction were made in 1945 when they tried after the Japanese collapse to set up communist administrations with the help of the war-time resistance forces, the Malayan People's Anti-Japanese Army. They had come out of the jungle and taken over control in many areas. However, they had to hand over when the British Military Administration arrived. When the communists started their campaign in 1948 they installed themselves in isolated communities of Chinese squatters, whose assistance in the form of supplies and recruits they obtained by terrorism and propaganda. They believed that the workers in the mines and on the estates would answer their call and join them there, where liberated areas could then be formed and extended.

---

[1] Mao Tse-Tung, *Aspects of China's Anti-Jap Struggle*, Bombay 1948, p. 68.
[2] For details about the organization of supplies, war industry and civil administration in Fidel Castro's war see Major E. Guevara, *La Guerra de las Guerrillas*, Havana 1960, Chapter III, 1, 2, and 6.

STRATEGIC ASPECTS

They did not succeed; their propaganda was ill-attuned to the temper of the population as a whole and the non-Chinese community in particular. Their task was not made easier by the Briggs Plan which, by denying them food supplies, forced large detachments to break up into smaller units, insufficient to establish a guerilla base; and General Templer's drive did not allow them to consolidate. Serious though the emergency was, the terrorists never rose above the roaming and raiding stage. They may have failed to learn Mao's lesson; they may have overestimated their appeal; most likely, they underestimated the tenacity of their opponent, who, without consciously preventing the formation of guerilla bases, denied the communists the opportunity for doing so. As it was, they fought a prolonged war without strategic means.

Mao's lesson was, however, not lost on the Viet-minh. They established their original bases in the north-east, near the Chinese border. One French fort after another in this region had fallen under their attack and they continuously extended their bases and created new ones. After the battle of Ninh Binh their army commander laid down that 'in order to attain our goal, we must achieve the following: develop guerilla warfare and strengthen local armies; destroy puppet administrations and extend the liberated areas in the enemy's rear; ... speed up the consolidation of people's bases....'[1] Mao could not have improved on that.

It does not so far appear to have been sufficiently recognized outside the circle of Mao's disciples that for revolutionary guerillas to succeed the formation of guerilla bases is of decisive strategic importance,[2] and conversely, it is of decisive strategic importance for anti-guerillas to deny guerilla bases to the enemy. When, in 1946, the Chinese communists resumed the civil war, they were too firmly entrenched over too wide an area for the Nationalist forces to do anything about it. But the war in Indo-China might have taken a different turn if it had been carried right from the start into the north-eastern region and the consolidation, if not the formation of the

---

[1] Quoted from Denis Warner, *Out of the Gun*, London etc., 1956, p. 151.

[2] As far as the author is aware, only Commandant J. Hogard, 'Guerre révolutionnaire ou Révolution dans l'art de la guerre,' in *Revue de Défense Nationale*, December 1956, recognizes the strategic role of the base, and so do implicitly 'l'Ecole d'Application de l'Infanterie de Saint-Maixent' in its *Manual* on anti-guerilla warfare, reprinted in *Bulletin Militaire*, December 1957, p. 857, and Ximenes, 'La guerre révolutionnaire et ses données fondamentales,' *Revue Militaire d'Information*, February/March 1957, p.19.

guerilla bases been prevented. We ought to know enough by now to assess their strategic importance.

It must not be forgotten, however, that it is only in a prolonged war that guerilla bases play a vital part in the strategic planning of revolutionary forces. If the revolutionaries can rally a large part of the population around them and if the opposing side, especially the regular army, is demoralized or in sympathy with the revolutionaries, the guerilla war is not a protracted war and bases, useful as they are, are not of primary importance. Castro's guerillas in Cuba had their base, but such bases are an assembly point rather than a nucleus State within a State—of organizational rather than of strategic value. But if the opponent is determined and well entrenched, the revolutionary war is of necessity prolonged. The protracted character of the war is turned by the revolutionaries into an asset: it wears down the enemy and gives the guerillas time to transform their inferior forces into superior ones. The base areas provide the means for doing that very thing.

According to Mao, the aim of guerilla war is to preserve and expand one's own side and annihilate or expel the enemy. The guerillas rely on their base area for carrying out their strategic tasks and attaining the war objectives in the four following ways:

1. The base fulfils the same function for the guerillas as the rear does for the regular army. In addition to its logistic role it is here that military power is consolidated and expanded; here guerillas and semi-regulars are recruited and trained.

2. The base is a communist-dominated enclave. Moreover, it is a political testing ground for the mass appeal of the communist programme. Here the communists mould the masses into organizations. They inquire into the opinions of the masses, 'co-ordinate' and 'systematize' these opinions and then explain and popularize them until the masses accept them as their own and translate them into action—a process known as the principle of 'from the masses to the masses'. If the mass response is insufficient, communist policies can—temporarily—be changed: we have already noted the switch in Chinese communist policy when they were preparing for the fight against Japan, and a no less significant change of course occurred in Malayan communist policies in 1951. In particular, the adoption of a 'reasonable' economic policy, even if it goes against the communist grain, is

considered of vital importance to the establishment of base areas. Thus political power is consolidated and the groundwork for communist expansion laid.

3. By expanding their own areas they deny the enemy more and more territory until he is finally expelled.

4. The last function of the base area is to serve in conjunction with other base areas and the semi-regular forces as the encirclement line of enemy held territory. While each isolated base area is encircled by the enemy, Mao explained,[1] neighbouring guerilla bases and the regular forces' front lines in turn surround the enemy. The encircling enemy thus faces encirclement himself.

Again, it was Mao who referred to the establishment of base areas in the anti-Japanese guerilla war as a *strategic* problem and *strategic* task and stressed the *strategic* role of established base areas.[2]

The anti-guerillas must therefore treat them as their *strategic* targets.

How, then, are base areas chosen? Not according to economic conditions, because base areas are established where the enemy is, and where he can sustain himself, the guerillas can too; 'where the enemy can go, there must for a long while have been Chinese inhabitants as well as an economic basis for making a living'.[3] Geographical suitability alone decides where to select bases. Base areas in mountain regions are the most desirable and those in river-lake-estuary regions are preferable to those in the plains, provided always that there is enough room for the guerillas to manoeuvre. Whether the base areas can form part of an encirclement line does not seem to influence the choice; they have in fact to be established 'wherever possible'.

In Yugoslavia Tito's partisan movement had recognized right from the start the need to form 'liberated areas'. As early as August 1941, a few weeks after the partisans had taken up the fight, they set up some sort of administration in the captured villages and small towns of Serbia, and a month later it was decided to extend the 'liberated areas' with communist-controlled National Liberation Committees as administrations. When the liberated areas fell to the enemy, new liberated areas were formed elsewhere, in Bosnia and Montenegro,

---

[1] In *Strategic Problems of Guerilla War* of which several translations exist. One is contained in *Selected Works of Mao Tse-tung*, vol. ii, London 1954. Cf. p. 144 for the above.

[2] *Ibid*, pp. 134, 140 and 145.

[3] *Ibid*, p. 142.

and by the second half of 1942 the partisans controlled large parts of Croatia and Bosnia; when Italy left the war, parts of Dalmatia and Slovenia were added, only to be soon lost again to the Germans. To the partisans the liberated areas were frequently a handicap. Their very existence offended against the classic guerilla rule of surprise; they invited attack by the enemy and often made the partisans lose the initiative; they restricted partisan elusiveness, and when the partisans had to give up liberated areas, part of the population left too, blocking escape routes and hampering mobility.

The Yugoslav partisans overcame the handicaps by good leadership and sheer bravery, but their experience underlines the validity of Mao's dictum that base areas must be vast to allow the partisans room for manoeuvre. 'In small countries, like Belgium,' Mao said, 'the possibility is very little or nil.'[1] If a country is not vast or large areas are not inaccessible, the strategic benefits of guerilla bases are largely cancelled out by the attendant tactical disadvantages.

The base areas can become a definite liability if for some reason it is imperative that the guerillas should hold them. Tito could relinquish base areas if he thought it advisable. Helped by the severity of German reprisals against the population, he could find popular support in other parts of the country. Arms and equipment he received by attacking the enemy and later the Allies dropped them where needed. He was not therefore dependent on a base area, least of all in his fight against Mihailovitch. For the Greek communist partisans it was however vital to keep control of a certain area and this led to their destruction.

At the end of the last war several thousand Greek communist partisans had sought refuge in Yugoslavia, Bulgaria and Albania; others went into hiding in the Greek mountains. These men formed the nucleus of the communist partisans who started the civil war in 1946. They occupied a number of areas in different parts of the country; the most important one for them was the northern area of Vitsi and Grammos on the borders with Yugoslavia and Albania from which countries they received support. This area they had to defend permanently to keep their supply and escape route open. In 1948 they had succeeded in holding out, after a temporary withdrawal into Albania, against vastly superior government forces; in 1949 they were decisively beaten: they had to give battle in orthodox war fashion for

[1] *Selected Works of Mao Tse-tung*, p. 142.

which they were not trained or equipped. Their positions were overrun and the war was over.

The moral is obviously this: revolutionary partisans must extend the areas of their political control and defeat the government forces. If the population does not overwhelmingly support the partisans and the government forces fight resolutely, the partisans must establish base areas. If the partisans receive logistic support from abroad through these base areas they must defend them to the last. They can only be defended, and the resistance of the government's forces can only be overcome, if the partisans have semi-regular forces at their disposal and apply orthodox military tactics. The Chinese and Vietminh had such forces and Tito was building them up; in Malaya and Greece there were none and in Malaya there were not even base areas.

Successful revolutionaries have always recognized the need for semi-regular forces in those circumstances. Guerillas can only weaken the opposing regular army and unless the revolutionaries can incite rebellion and revolt they must defeat it in the field, with their own regular or semi-regular forces. The obvious way for revolutionaries to obtain regulars is to raise some of their guerilla units to the regulars' standard. This is what General Galen did during the Russian civil war. He was later a member of the Borodin mission to China and Mao in his *Anti-Jap Struggle* refers approvingly to Galen's achievement. Mao learned from Galen and Ho from Mao; in their wars they employed armies composed of semi-regulars and guerillas. Thus 'the strategic role of [revolutionary] guerilla warfare is twofold: supporting regular warfare and transforming itself into regular warfare'.[1]

During the anti-Japanese war Mao's regulars were not much in evidence. There was, of course, the Eighth Route Army consisting of three divisions. It had been built up in the time before the Japanese invasion from a guerilla force into a somewhat more formal army, only to revert again to guerilla warfare during the war against Japan. Mao foresaw the need for this army to change over to regular warfare. Yet for the time being he considered a division of labour 'necessary and proper ... with regular warfare carried on by the Kuomintang on the main front and guerilla warfare carried on by the Communist Party in the enemy's rear',[2] an arrangement which, as he was

[1] Mao Tse-tung, 'On the Protracted War,' in *Selected Works*, vol. ii, p. 224.
[2] 'Problems of War and Strategy.' Speech delivered at the Central Committee's plenary session in November 1938, reprinted in *Selected Works*, vol. ii, p. 279.

quick to point out, had the advantage of allowing the expansion of the base areas, the Communist Party, and the forces. By the time the civil war was resumed the communist forces had once again become semi-regular troops, supported by guerillas.

How is this transformation into semi-regular status effected?

When the Chinese 'Workers' and Peasants' Army' was formed in 1927 and when Tito organized his partisans into a 'National Liberation Army' in 1942, they were armies in name only. They lacked the necessary arms and equipment and for this reason alone their actions were almost exclusively of the guerilla type. They could engage in more conventional combat only after they had made good their deficiencies; the Chinese had to wait for their chance until 1945 when they seized large Japanese stores and, from 1946 on, captured U.S. *matériel* from the Nationalists, while the Yugoslavs had their first stroke of luck when Italy surrendered in 1943 and they got hold of sizeable portions of Italian arms and equipment; later, as we have said, they received Allied aid on a substantial scale.

Even so, the revolutionaries were almost exclusively infantrymen—they had practically no air force, very few tanks, no medium or heavy artillery—and their guerilla experience stood them in good stead: guerilla discipline had been strict, they had learned marching in their guerilla exploits, as guerillas they had known how to use what weapons they had had, and they had all been in battle. What more there was to know they mostly acquired as they went along; where possible they were retrained after battle, in the light of experience; in particular the Chinese made good use in this respect of the 1946 truce period in Manchuria.

For weapon training both Yugoslavs and Chinese had outside help. The British mission to Tito included artillery instructors and an artillery school under British command instructed the partisans in the use of pack-howitzers, while the Chinese communists made use of Nationalist regulars and Japanese technicians and mechanics.

Almost from the start, the Yugoslavs and Chinese had built up their forces organizationally from lower to higher formations, Mao starting with an 'army' of 2,000 men and Tito, more modestly, with brigades of almost equal size. Both had established a chain of command. The leadership of the party was never in doubt. In Yugoslavia, where there was only guerilla fighting until Belgrade, the battle-tested guerilla leaders proved equal to the task of commanding partisan formations which were growing ever larger by the intake of new recruits—an entire division of 3,500 men would hold a German advance, secure a flank or force a river—and they had therefore had experience in handling larger formations when warfare became more conventional, at and after Belgrade. In China a surprising number of communist generals were regular soldiers who had graduated from Chinese national or Soviet Russian military academies. Chinese and Yugoslavs alike relied for food on the land and for arms and equipment on their enemies—and Yugoslavia also on the Allies. The political work was in the hands of the commissars and political cadres. Apart from training, the higher commanders' tasks were therefore almost entirely operational and they were simplified because in China life is expendable.

Other officers and NCO's were either party men or had been promoted from the ranks or come over from the enemy—Chiang and Mihailovitch. Chinese junior officers were also trained at the Military and Political College in Yenan.

What the officers lacked in professional knowledge and the men in training they learned the hard way in combat. That this system worked is partly due to the fact that the change-over to a more formal army was gradual. In China from 1938, and in Yugoslavia throughout the war, losses in men could always be more than made good. What mattered most was that morale was never affected by the losses; owing to ceaseless political propaganda it constantly remained high.

The transformation into semi-regular forces must be properly timed. Semi-regulars can, of course, always revert to guerilla fighting, but if they intend to fight orthodox battles they must wait until they are strong enough to do so, and that means until they are not greatly inferior to the opposing forces. Tito rightly bided his time until the Red Army had entered Yugoslavia and was able to assist him in the field; his attack on Belgrade was strongly supported by Soviet tanks. Mao, in the resumed civil war, did not discard guerilla methods until the disparity in numbers and *matériel* had been considerably reduced, and even then the guerillas remained in being as a separate force. However, the post-war Greek guerillas, as we have said, did not have semi-regulars, but they were on the point of building up such a force; in 1948 their detachments were first grouped into brigades and then into divisions. But their total strength was only one-tenth of that of

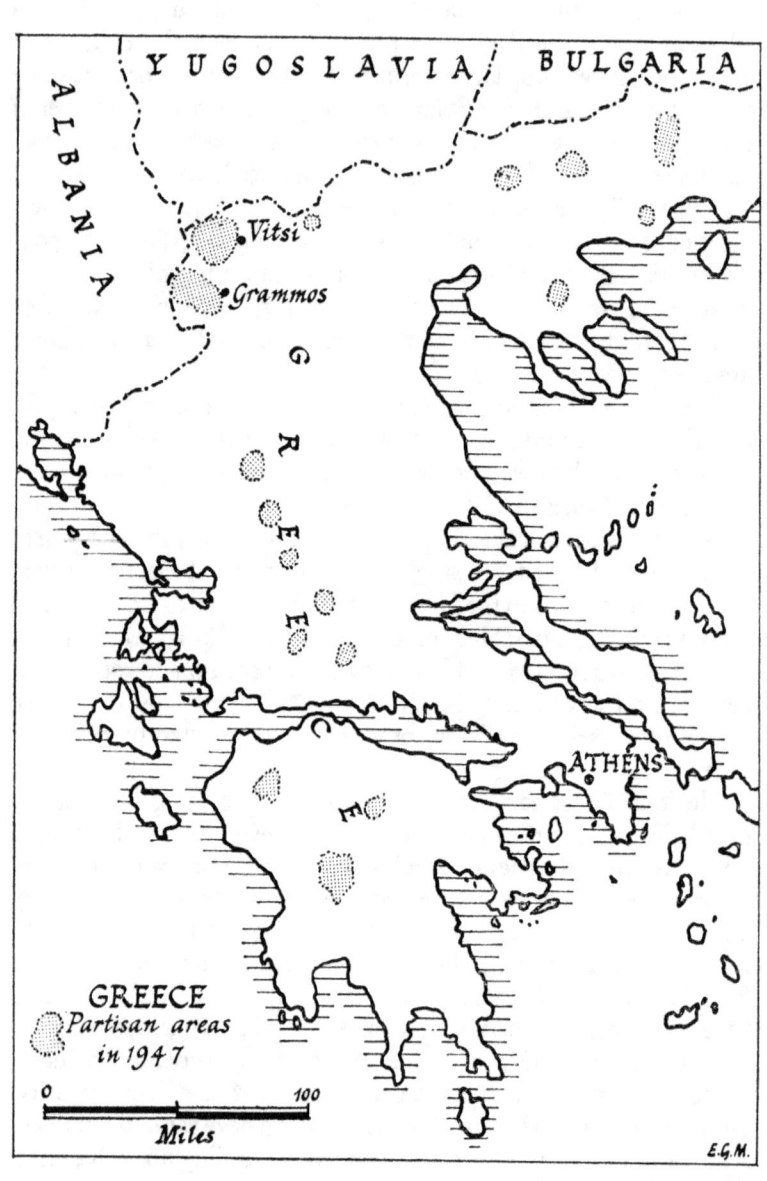

the government forces and the eventual outcome of the campaign was a foregone conclusion when, in the attack on their vital base at the Yugoslav-Albanian border, they had no choice but to defend it in orthodox fashion.

If timing is important for the guerillas, it is no less important for the anti-guerillas. For the anti-guerillas vital base areas are not only a strategic target but also the top priority target. This target must be attacked before the guerillas have become strong enough to defend it successfully on conventional lines. If the base areas are outside the country it may not be possible, for political reasons, to attack them at all, but if they are in the country and the government forces are superior they must attack them at the very first possible moment. An attack on vital base areas under such conditions is the only strategy which the anti-guerillas can employ in order to make the guerillas face a superior regular force in conventional combat while denying them all guerilla advantages.

The guerillas must try to prevent their opponent from building up a superior force; by continuous attacks on enemy-held towns and villages, his lines of communication and stores, over as wide an area as possible, they will try to make him disperse his forces on security duties and also gain time for their own build-up. These tactics were adopted by the Greek communists and they were successful until the Greek Army formed national defence corps and home guards, trained by the military, for local self-defence, thus freeing the regular troops hitherto employed for this purpose. Wherever this is feasible it must be done; and if it is not possible because the population sides with the guerillas, a calculated risk must be taken and the anti-guerillas must denude the country of security forces, if with their help they can build up a superior force for the attack on the vital base. It must be borne in mind that it took the Greek Army only just over a week to take the vital communist base area and achieve the decisive victory in a war which had dragged on for three years.

If the anti-guerillas have superior forces it would be a strategic blunder to cut the guerilla lines of logistic support instead of attacking the vital base. Once the lifeline is cut and cannot be restored, the base ceases to be vital for the guerillas and they have no reason to defend it in the orthodox way. They will disperse and restrict themselves to harassing actions which may continue for years and ruin the country's economy. They may also try to form new bases. The guerilla war, which would have collapsed with a successful attack on the vital base, goes on.

The aim of the attack on the base area is the elimination of the guerillas. Their escape route to the border must be closed. The anti-guerillas' battle technique is that of encirclement and annihilation.

Under the same conditions as those indicated here for revolutionary guerillas, all independent guerilla movements fighting with the aim of defeating the opposing army in the field must have base areas and acquire a semi-regular army, and the foregoing rules for guerilla and anti-guerilla fighting apply to them as well.

Guerillas may of course try to reduce the importance to them of the base area, but there is little they can do. After all, they must have, hold, and expand their area and progressively deny ground to the enemy. They may, however, try to make themselves less dependent on the base area for logistic support: in the Greek civil war 1947–49, instead of delivering all the supplies to the frontier and the adjacent vital base area, the Yugoslavs could have supplied each of the guerilla areas directly by air-drops. While this did not happen then, it might happen somewhere else on a future occasion.

We have dealt here with strategic aspects of guerilla warfare and we now propose to look at some operational problems.

CHAPTER 4

THE CHANCES OF SUCCESS: OPERATIONAL ASPECTS

It has been repeatedly pointed out in reviews of the Chinese war 1946-49 that while the guerillas and semi-regulars excelled in mobility their opponents increasingly lost their offensive spirit and developed what has been called a 'pillbox psychology'.[1] These factors, it is held, greatly influenced the outcome of the war. Exactly the same observations have been made in regard to the Indo-Chinese war. It therefore seems opportune to inquire whether guerilla tactics in both wars forced the opponents into a state of immobility and static defence and, if so, how this was achieved.

At the end of the war against Japan the Chinese communists occupied sixteen base areas with a population of about 100 million people. Twelve of these areas, the most important ones, were located in northern China in a huge triangle with Yenan, the communist capital, as apex and the coastal provinces south of Peking and well north of Shanghai as base. Since China's main railway lines from north to south run through this region it was ideally situated for cutting communications and impeding the enemy's troop and supply movements. The communists' semi-regulars numbered about 900,000 and they also had militia and guerillas, while Chiang had 2,500,000 regulars under arms, reinforced by 1,500,000 men under their warlords. A considerable part of Chiang's force was American trained and equipped. The nationalist soldiers were badly fed and badly paid and, with the Japanese invader gone, they hardly knew what they were fighting for. They had been at war for years and the seeming purposelessness of this new war had a stronger effect on their morale the longer the conflict lasted. On the communist side the soldiers knew their war aims; indoctrination had seen to that; and they could observe for themselves the reformation of community life in the villages and fields.

[1] Colonel Robert B. Rigg in *Red China's Fighting Hordes*, Harrisburg 1952.

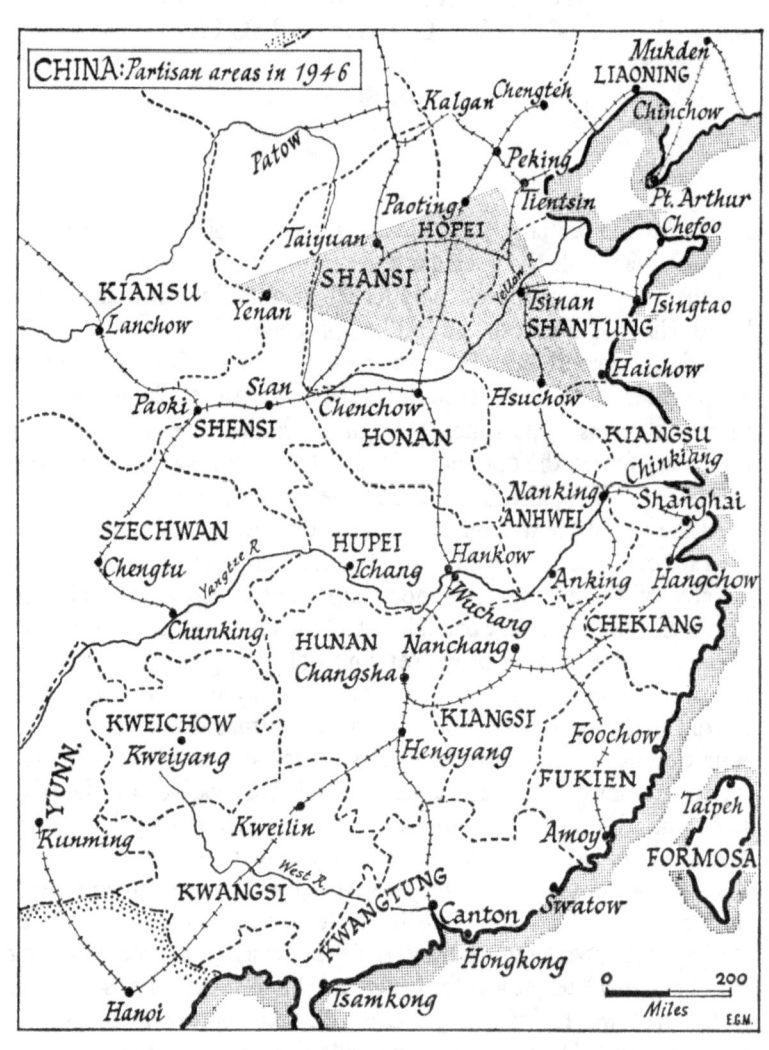

OPERATIONAL ASPECTS

The nationalists' stronghold was in the south, in the Yunnan and Szechwan provinces with their capital Chungking, where both sides met to discuss the communist share of the future Army and the fate of the communist liberated areas. No agreement could be reached and hostilities broke out. The nationalists had two advantages over the communists: the Japanese had been given orders by General MacArthur to surrender only to nationalists and the communists were duly prevented by the Japanese from occupying the cities they had held; furthermore, the nationalists were airlifted by the Americans into these cities; Peking, Tientsin, Nanking, Hankow and Shanghai were thus all taken over by the nationalists.

There was little the communists could do about this—apart from cutting the north-south communications—since U.S. marines had come to the key cities in order to arrange for the return transport of the Japanese troops, but Manchuria held out better prospects, since the Russians were here in control. Here were not only heavy industry and coal and iron mines but also large Japanese arsenals which the communists urgently required. The nationalists had equally pressing motives for entering Manchuria; above all they had to assert China's sovereign rights over the former Japanese puppet State of Manchukuo. Again, the nationalist troops were airlifted and took possession of Mukden in the south, while the communists marched and occupied the north and west.

In January 1946, the nationalists and communists came to an agreement about their differences, but the terms were not observed by either side. In April 1946, the civil war broke out[1] and the communists took possession of the rest of Manchuria without firing a shot. However, with their heavy arms and better equipment the nationalists soon managed to drive the communists out of the capital Chang Chun and other cities and towns.

In North China the communists were now marching towards Peking and had encircled Tsinan, but the nationalists had local successes in the communist strongholds, the provinces of Shantung, Hopei and Kiangsu, as well as against smaller forces in the centre and south, which then joined up with their main forces in the triangle in the north, leaving guerillas behind. The communists thus had their

[1] For the history of the war cf. Général L.-M. Chassin, *La Conquête de la Chine par Mao Tse-tung* (1945-1949), Paris 1952, on whom we rely here for the factual description of the campaign.

troops concentrated in two main areas, North China and Manchuria, while the nationalists were dispersed all over the country.

This first phase of the fighting was on a small scale and on the communist side it was only guerilla warfare. Each side had tried to secure for itself the larger part of the areas evacuated by the Japanese, but in the process the communists had been more selective. The nationalists, as the lawful government, had to assert their authority all over the country. By contrast, the communists were not yet interested in central and southern China, apart from fomenting guerilla activities there. In other words, the nationalists' decision about the disposition of their troops was influenced by political considerations, while the communists acted on purely military ones. The result was that Mao, in spite of his over-all inferiority, could now hope to mass locally superior forces against the enemy.

This was a factor of the utmost importance in his plans. As he puts it, he could in future employ 'numerically preponderant forces to launch attacks of quick decision from exterior lines so as to capture a greater number of enemy soldiers and a larger quantity of spoils than would otherwise be possible'. 'To carry out this directive', he goes on, 'we need initiative.' 'By the initiative we here mean an army's freedom of action as distinct from a state of passivity into which an army may be forced. Freedom of action is the very life of an army and once this freedom is lost, an army faces defeat or annihilation.... We can say that the exterior-line quick-decision attacks, which we have proposed, as well as flexibility and planning in carrying out such attacks, tend to gain the initiative in order to force the enemy into a passive position.... Initiative or passivity is inseparable from superiority or inferiority in fighting strength. And consequently it is also inseparable from correctness or incorrectness in command.'

'Passivity', Mao concludes, 'is ... always disadvantageous and one must try to get out of it by all means. The way to achieve this militarily is resolutely to launch exterior-line quick-decision attacks and to start guerilla warfare in the enemy's rear to secure partial yet overwhelming superiority and initiative over the enemy in numerous campaigns of mobile and guerilla warfare.'[1]

It is therefore obvious that Mao intended by his conduct of semi-regular and guerilla warfare to force his enemy into passivity.

In the second phase of the operations, which covers the year 1947,

[1] Mao Tse-tung, *On the Protracted War*, loc. cit., pp. 211 f.

the communists put their military doctrine into practice. At the beginning of the year they were still greatly inferior to their opponent; they had 1,100,000 men against 2,600,000. The nationalist forces in Manchuria and North China depended on Central China for supplies. By the end of the year the communist had pinned down the nationalists in these areas by cutting their lines of communication; they then attacked with locally superior forces one isolated defence pocket after the other.[1]

The communists were particularly successful in Manchuria. In five offensives during the first half of the year they succeeded in encircling the principal nationalist garrison towns, Mukden, Chang Chun and Kirin, and took many others. In spite of counter-attacks, the nationalists were unable to restore the situation, and by the middle of the year they had lost half their Manchurian territory, two thirds of their railway lines and great quantities of arms and equipment. Chiang sent his best troops as reinforcements. In the second half of the year the communists mounted two more offensives in Manchuria and the nationalists were compressed into a few cities and strongholds. Their attempts to reopen the blocked railway line between Peking and Mukden remained fruitless. In North China the communists gained control of practically the whole of Shansi province, they occupied Hopei and established a base in Honan province, thus extending the area of their original bases beyond the triangle to the north and south.

During the first half of the year the nationalists had certainly not yet lost the offensive spirit. They had counter-attacked in Manchuria and they had also attacked in Shantung, Hopei and Shansi provinces. But what gains they had were dearly bought and, because of enemy guerillas, militia and newly formed semi-regular forces, they could not pacify the area they had taken.

There now came the first reports that many nationalist soldiers had deserted and that morale among the troops was generally low. They were also said to be developing a defensive mentality. There were a number of reasons for this. Mao's forces were mostly locally raised and were defending their homes, while the nationalists, coming from all parts of China, fought on almost alien soil. Mao's troops seemed to have paid for what they received from the population, while the

[1] Cf. *The China White Paper*, a Summary with Commentary of the Department of State's 'U.S. Relations with China', by Francis Valeo, Washington 1949, p. 40.

underpaid nationalists pillaged what they needed; the population therefore resented their presence. But there were other reasons too, and for these Mao can take full credit: the nationalists had lost one engagement after another, they had been confined to isolated outposts, they had been unable to stop guerilla attacks on the railway lines or on themselves and they now kept to their blockhouses. 'The extensive development of our guerilla warfare in his rear', says Mao of his former Japanese enemy, 'has driven his garrisons in the occupied territories into a completely passive position.'[1]

The communists had managed during the year to reduce the disparity in numbers and they then started to discard the guerilla character of their operations. Three factors helped them to make the disparity now less significant. In the first place, when they had taken an area they could withdraw their troops, while the nationalists were handicapped by attacks from the guerillas and the hostility of the population and had to keep the area occupied.[2] Secondly, the nationalists put garrison troops into every town and thus split up their forces, and thirdly, they adhered to the maxim that a three-to-one superiority was needed for attack and they were now hardly able to concentrate such superior forces. From now on the strategic initiative was Mao's.

This became increasingly obvious during the third phase of the campaign, which started in 1948. The communists suffered some setbacks but they more than made them good by immediate attacks elsewhere. The few remaining nationalist strongholds in Manchuria were still further reduced. In South China guerillas became increasingly active, especially round Hongkong and Canton and along the Indo-Chinese border, tying down a considerable number of nationalist troops. In North China the communists re-took Yenan, which they had evacuated under nationalist pressure the year before, yet immediately afterwards the communist forces under Peng Teh-Huai, menacing Szechwan province, lost half their effectives in battle. However, they managed to hold on to Shensi and Shansi which they soon completely controlled. The guerillas became even more aggressive everywhere, they took one nationalist fort after the other and enforced a blockade on Tsinan and Tsingtao. In the battle of Kaifeng, near Chenchow, the communists fought for the first time in the conventional way, with massed infantry, artillery and even a number of

[1] In *On the Protracted War*, loc. cit., p. 218.   [2] Général Chassin, loc. cit., p. 154.

OPERATIONAL ASPECTS

tanks.[1] They took the town but evacuated it again. By the middle of the year the nationalists held on to the Peking-Tientsin-Mukden-Chengteh sector in the north and the Hsuchow-Chenchow-Sian sector in the centre.[2] They soon lost Mukden and with it Manchuria; its defence had cost the nationalists at least 400,000 men[3] while the communists were free to engage their forces elsewhere. They gained control of Shantung, Shansi and West Honan. Central China now became the principal theatre of operations.

The communists took Hsuchow and moved on to Nanking. The forces on both sides were equal in numbers. The nationalists waited for the enemy to attack. Hampered by desertion, as they had been before in Manchuria and at Tsinan, they were forced to withdraw their defences outside the town and then decided to evacuate it, but the communists succeeded in encircling most of the garrison. Many went over to the communists, the rest were taken prisoner or fell in battle. The nationalists had lost another 600,000 men.[4]

When war broke out the nationalists had had an almost fivefold superiority in numbers; now the communists were superior. At the beginning of 1949 the nationalists had 1,500,000 men and the communists over 1,600,000.[5] The nationalists decided that each governor was from now on responsible for the defence of his province and 'this meant in practice that co-ordinated resistance had ended'.[6] Peking, long isolated, surrendered in January and Chiang departed. On April 20, 1949, the communists began the final offensive across the Yangtze river. By December the war was over.

The best postscript on the campaign was written by Mao long before the event, in May 1938, when he outlined his ideas on the conduct of the war against Japan. 'War is a contest of strength, but the original state of strength changes in the course of war. Here the efforts one makes to score more victories and commit fewer errors are the decisive factor.... While we strive for superiority and initiative the enemy strives for them too; thus, war is in truth a contest in ability between the commanders of opposing armies in their struggle for superiority and the initiative.'[7] A review of the Chinese Civil War shows that in this struggle for superiority and initiative the semi-

---

[1] Général Chassin, *op. cit., p.* 165.
[2] Général Chassin, *op. cit.,* p. 173.
[3] Général Chassin, *op. cit.,* p. 183.
[4] Général Chassin, *op. cit.,* p. 190.
[5] *China White Paper, op. cit.,* p. 40.
[6] Général Chassin, *op' cit.,* p. 197.
[7] In *On the Protracted War, loc. cit.,* pp. 211, 214.

regulars and guerillas had certain advantages which the communists consciously exploited.

They had of course had long experience in warfare and it is therefore not surprising that Ho Chi-minh, who had only a few years of resistance against the Japanese to his credit, failed at first to follow these rules in the Indo-Chinese war. Before the French, with British assistance, had returned to Indo-China after VJ-day, Ho had already proclaimed the Republic of Viet-minh and become its first president.[1] Hostilities between his troops and the French broke out before the end of 1945. He had given his early attention to forming an army and by 1946 he had over 30,000 'regulars' at his disposal. Their main task was the defence of the two principal base areas in the highlands, while regional battalions and village self-defence units, acting independently and decentralized, had to harass the French in the plains; when these latter forces became proficient they could be incorporated into the regular army, and only the best were actually taken over. Some Japanese technicians helped with this army's early training, and its arms were Japanese, American, Chinese and French.[2]

In December 1946, the Viet-minh rose against the French all over the country, and they launched many guerilla operations in the following months and years. Only militia and self-defence units took part in the early activities, and the encirclement in 1947 of French garrisons in Hanoi, Haiphong, Nam Dinh and elsewhere was carried out without the support of regulars. The French, reinforced from other parts of the country, soon broke the encirclement. Nor were there any regulars involved when the French started later in the year to sweep the country north and south of the Red River, and resistance was weak. However, the guerillas were already causing them considerable trouble and although they managed to keep some of the roads open they were unable to pacify the country.

During this first phase all operations were of the guerilla type and they attracted little attention outside the countries involved. This changed when it became clear that the Chinese communists, after their victory over the nationalists, were supporting the Viet-minh by

[1] Ho had previously worked under Borodin in Canton and subsequently helped to form the Malayan Communist Party.

[2] The Japanese arms had been taken from them during their occupation of the country in World War II, the American arms had been dropped to the guerillas also during the last war, the Chinese arms had been bought from Civil War contestants, and the French arms came from former arsenals.

training and supplying them with arms and equipment. In the autumn of 1950 the Viet-minh regular forces of thirty battalions took the offensive and, catching the French unawares, forced them to evacuate the mountain region of Tongking and the posts in the Lao Kai area, on the Red River near the Chinese border, at Kaobang, and at Langson, also on the border further to the east. But the French rallied and their opponents were denied possession of the Red River delta. The Viet-minh realized that they could not yet win in open battle and from now on they co-ordinated the operations of regulars and guerillas. Thus the second phase of the war began.

Most of the subsequent fighting took place in Tongking but there were also encounters in Cochin-China and elsewhere. Tongking, particularly the delta, was the most important theatre of war. For Ho it meant recruits, rice, and supplies by sea from China. Although he failed to gain possession of the ports, he succeeded in securing for himself the manpower and supplies of the area, and the French defence system helped him to do so. The French had put their trust in strongpoints and earthworks. The Viet-minh regulars usually attacked a fort in the delta in order to distract attention from their guerillas, who would then slip through to the French rear. They enlisted the population of certain villages as their intelligence agents, designated other villages as defence posts and there raised regional battalions and guerillas to operate locally by harassing enemy patrols and attacking vulnerable forts.[1] The Viet-minh thus hoped to drain the enemy's strength. When the guerillas went back to their own area, they brought not only supplies but also recruits. The French were unable to establish a front in the delta.

Neither side was strong enough to drive the enemy from the delta. The French hoped for considerable reinforcements from the new Vietnam Army, while Ho's forces became increasingly better trained[2] and armed. The year 1951 did not see any decisive battles, but Ho maintained constant pressure in Tongking and the south, though often at considerable cost to himself.[3] The pressure became so strong in Tongking that the French decided, early in 1952, to withdraw their garrison of 20,000 from Hoa-Binh on the Black River and from Colonial Highway No. 6, its link with the delta. In the delta itself the

[1] They were trained by infiltrated military instructors who also saw to it that the village was fortified and that arms were produced locally.
[2] His gunners received their training in Kwangsi in China.
[3] Cf. Bernard B. Fall's brilliant *Street Without Joy*, Harrisburg 1961, pp. 27 f.

OPERATIONAL ASPECTS

Viet-minh extended their foothold; they now had more than 20,000 men under arms there, over half of them regulars who had infiltrated and were stationed in the defence villages.

In October 1952, 30,000 Viet-minh regulars, supported by regional battalions, started their offensive in the direction of the Thai country. Their first objective was Nghia-Lo in north-west Tongking. The French Union forces lost the town, then withdrew from Gia-Hoi, south of the Red River, and the Viet-minh continued their advance towards the Black River, which they crossed. The French Union forces tried in the meanwhile to cut communications with China, and to extend their hold on the delta by means of airborne operations. Again and again the Viet-minh, in combined operations of regulars and guerillas, attacked French Union posts on the 165-mile long defence line from south Laichau province of Mocchau in Son-La province; sixty-five posts fells to them in quick succession, until only Na-Sam held out.

In April 1953 the Viet-minh invaded Laos with the intention of drawing French Union troops from the delta, and then withdrew most of their forces again. Some of their units fought their way into the delta and joined those already there.

By now the Viet-minh had a standing army of fifty battalions, a local people's army consisting of 30 regional battalions, and militia and guerillas, capable of launching an attack almost anywhere. The French wanted to counter this threat with a mobile reserve, including airborne units, which could be deployed as an emergency defence force wherever needed, but manpower on a sufficiently large scale was not available, as there were doubts about the reliability and fighting qualities of the new Vietnam army. This did not prevent the French from making parachute and commando raids or bringing up reinforcements by air into the threatened areas, but their freedom of action was limited. They were, in particular, unable to stop further infiltration into the delta, and by December 1953 the Viet-minh had increased their forces there to 60,000 men. The French evacuated Laichau in order to seize and fortify Dien Bien Phu as a base for operations to clear the Thai country of the enemy and block a gateway to Laos. The Viet-minh chose another entrance; while they bypassed the French Union Forces in Dien Bien Phu, 20,000 of their men, taken from the Hanoi area, cut Indo-China into two by a lightning thrust from their coastal base at Vinh through Vietnam and the entire width of Laos; they reached the Siamese border, but the French stopped their further advance.

In January 1954, the Viet-minh stepped up operations, including guerilla activities, in the three theatres of war—in Laos, round Dien Bien Phu, and in the delta—and succeeded in keeping the French Union forces dispersed, tied down to their defence posts and unable to gain the initiative. Then the Viet-minh opened the attack on Dien Bien Phu with regular forces while guerillas kept the Union forces in the delta in a constant state of alert. In spite of parachute reinforcements the French garrison was unable to stem the Viet-minh advance on Dien Bien Phu, which fell, after heroic resistance, in May 1954.

While world attention was still focused on Dien Bien Phu the Viet-minh took the offensive again; 30,000 men marched on the delta, took a strategic town near Hanoi and pressed on. In June the French Union forces evacuated the southern zone of the delta. This little-known operation was indeed the greatest of all Viet-minh victories in this war. By combined operations of regulars and guerillas they had succeeded in completely immobilizing the Union forces in their positions; some of the southern delta regions had been infiltrated to such an extent that the entire area had become untenable for the French Union forces. In 1950 the Viet-minh had begun this process of infiltration and in four years they had built up their strength until their opponent was forced into passivity and withdrawal. As a military operation it had brilliantly succeeded. General Giap, the Viet-minh C-in-C, had shown himself a truly outstanding disciple of Mao.

By then the war was practically at an end and the tactics of the delta were not repeated elsewhere. The Viet-minh were intensifying their campaign round Hanoi and in the centre and south of the country when the truce, soon followed by the Geneva Agreement and the division of the country, brought fighting to an end.

It may be argued that the immobilization of the opponent is the automatic and inevitable by-product of any war fought by regulars with guerilla support, at least if the opponent is numerically inferior. Looking back, it is true that the nationalists in China and the French in Indo-China were trying to cover the entire territory with their forces, and this deployment made it difficult for them to attack.[1] It is

---

[1] For a criticism of the French conduct of the war cf. Amiral Castex, 'Les enseignements de la Guerre d'Indochine', in *Revue de Défense Nationale* 1955, vol. xxi, pp. 523 seq., and Bernard B. Fall, *op. cit.*, especially pp. 159 f. and 223 f.

OPERATIONAL ASPECTS

also true that in order to protect, the protective forces must stay put in the places they are guarding. But it is wrong to conclude that they must therefore develop a Maginot line spirit. After all, the German anti-guerillas in Russia, although locally outnumbered, never allowed themselves to be forced into passivity. The German directives in *Warfare against Bands* stated specifically that 'in the fight against bands the initiative must always rest with us. Even if the commander has only a small force at his disposal he must not fail to show resolution. If possible, each action by bands must be followed by counter-action.' In particular, by forming *Jagdkommandos* (commando hunters) of platoon or company strength, the Germans found it possible 'to fight actively against the bands even with the smallest forces'. And the commanders of troops on guard duties were instructed to deploy as guards only as many men as were absolutely necessary to fulfil their protective tasks; all other forces at their disposal were to be used for combating the bands. These directives were carried out in practice.

In their anti-partisan operations the Germans were hampered by local inferiority in the number, and often in the quality, of their troops, which consisted of many elements: front line troops, their own security and police units, allied security units, and a mixed bag of native security and defence forces. Unable to hold the entire countryside, the Germans wisely restricted themselves to maintaining garrisons in selected localities and guarding vital installations and communications. A post along the lines of communication was hardly ever manned by more than fifteen soldiers, but even so four were deemed sufficient to defend it while the others moved out on nightly patrols and fought against partisans in neighbouring villages, along the line, or in support of adjoining posts.[1] Again, when in a sector new native units had been recruited or security troops could be spared, even temporarily and in small numbers, they were employed as reconnaissance troops to seek out and fight the partians.[2] Side by side with these small actions went many on a medium scale and when front line troops were available, large-scale operations for combing out an area and encircling the partisans were mounted, one in June 1942 in the Bryansk area, five a year later in the same area and one in the Vitebsk area, and three more in the first half of 1944 near Orsha-Vitebsk and Minsk.[3] German successes in these operations were

[1] Cf. Walter Hawemann, *Achtung! Partisanen*. Hanover 1953, p. 22 f.
[2] Major Howell, *op. cit.*, p. 178.   [3] *Ibid*, pp. 94, 158 f., 200 f.

limited, but the Germans made the most of the little they had and at no time did their anti-partisans become paralysed. The same holds true of the German anti-partisans in Yugoslavia.

It is therefore obvious that, even if an opponent is weaker, guerillas can force him into passivity only if he disperses his forces over too wide an area, deprives himself of the means of attack and thereby loses the will. Since he cannot then protect the population, he is almost bound to lose the war.

There are two further lessons to be learnt from the foregoing.

First, the anti-guerillas must avoid being encircled, as the Chinese nationalists were in many places and the French Union forces were at Dien Bien Phu. Colonel Rigg was a U.S. observer at the battle of Chang Chun in Manchuria. Chang Chun, like Dien Bien Phu, was air-supplied and lost because of concentric envelopments by the opponent. Colonel Rigg came to the conclusion that in order to survive the airhead must move and keep on moving on the ground in order to force the enemy to regroup his troops and prevent him from concentrating them.[1] Moving the garrison might create new supply problems, but unless the town can be relieved Colonel Rigg's idea seems to be the right one. If it is accepted it follows that reinforcements should not be dropped into the beleaguered garrison, but outside, so that they can create a diversion.

Secondly, it would be of great advantage to the anti-guerillas if they in their turn could fight their opponent to a standstill and into passivity. But unless he has a vital base area to guard, no simple means similar to those open to the guerillas are available to the anti-guerillas. Partisans, after all, have little or nothing to guard. If their camp is threatened they give it up rather than fight against odds, and their supply lines, if they have any at all, need little or no protection. For the supply of arms and equipment the great guerilla movements of the recent past were usually not, or only to a small extent, dependent on vulnerable road and rail communications: what the Soviet and Yugoslav partisans needed was either taken from the enemy or supplied by air from Allied sources, and for food they relied on the land. So did the Chinese and they supplemented captured arms by local production. The post-war Greek guerillas transported their supplies by packhorses and mules; only the Indo-Chinese used lorries for the transport of arms and even their demands on road transport

[1] In 'Red Parallel, The Tactics of Ho and Mao', *Army Combat Forces Journal* 1955, p.30.

OPERATIONAL ASPECTS

were small by comparison with army requirements; an estimated 75,000 coolies carried most of the load and they kept to the tracks.

Apart from this, guerilla actions are based on surprise, quick decision and mobility in attack and dispersal, and these tactics do not allow of passivity. Equally, Mao's famous Ten Principles for the conduct of guerilla warfare, expounded in his speech to the Central Committee of the Chinese Communist Party of December 25, 1947, are so designed as to keep the guerillas in an almost constant state of movement. 'Do not let the enemy have breathing space', 'Train to fight successive engagements within a short period', 'Strike first at scattered and isolated enemies', in fact his emphasis on taking first what is easy to take—'Take first the small towns', 'Take first weakly defended cities'—and on the need 'to destroy the enemy while he is moving', of necessity keeps the guerillas on the move themselves.

It taxes the ingenuity of the anti-guerilla commander to deny his enemy room for manoeuvre and deprive him of the will to fight. Yet once the problem is recognized a way towards its solution can be found. This was shown in post-war Malaya by General Briggs and his successor, General Templer.

During the first two or three years of the emergency in Malaya the British attempted to penetrate the jungle in battalion strength, but the guerillas always had early warning of their approach and the missions accomplished little. The initiative usually rested with the guerillas, who managed to inflict considerable damage. It therefore became clear that the British required considerably more knowledge of the ways of guerillas before they could hope to beat them, and for this purpose General Briggs set up a 'ferret force' of British, Malayan, Gurkha and Chinese troops. This force lived in the jungle for months in order to carry out patrols and obtain intelligence on the MRLA. After its return members of the force were assigned to various units as information and intelligence officers so that these units could benefit from their experience. From then on the British sent only squads and platoons into the jungle. They copied the techniques of the guerillas, they became adept at avoiding detection and highly skilled in ambushing and cutting supply lines. Like guerillas, they lived on the land and frequently changed their location.[1]

[1] Cf. for the above in particular James E. Dougherty, 'The Guerilla War in Malaya', *U.S. Naval Institute Proceedings*, vol. 84, No. 9 (September 1958), pp. 41 seq., and also Lt.-Col. Joseph P. Kutger, 'Irregular Warfare in Transition', *Military Affairs*, vol. xxiv, No. 3 (1960). pp 121 f.

PARTISAN WARFARE

Hand in hand with these new tactics went the food denial measures, the removal of the jungle fringe squatters who had supplied the guerillas with food; their resettlement in areas where they could be protected against guerilla raids and enjoy for the first time electric light, water, good roads and security of land tenure; and the formation of the aborigine fighting regiment which has already been mentioned in Chapter 2. These measures, in combination, had a double effect: they won over a large part of the population to the British side and they led to an extension of the guerilla supply lines. Since these could now be ambushed the guerillas had less and less contact with the population and also had to grow their own food in jungle clearings. These clearings provided a good target for the RAF and the guerillas were gradually driven deeper and deeper into the jungle. They had become isolated.

Neither resettlement nor the employment of a ferret force was a new idea, but by their ingenious combination the British achieved success. The Germans in Russia had evacuated the population from guerilla areas—without, however, gaining its allegiance—and they had employed *Jagdkommandos* in the same areas. The *Jagdkommandos* were first formed in 1942. As we mentioned earlier in this Chapter they were of company or platoon strength and their purpose was to make possible attacks on bands even when only small forces were available. The leading idea behind their battle technique was unobtrusively to get as near to the bands as possible by imitating their fighting technique and assimilating themselves to local conditions. They then had to annihilate the bands by surprise action. The *Jagdkommandos* were in fact shock troops, and in all these respects the Malayan antiguerillas resembled them.

The mission of the *Jagdkommandos* was to see to it that the bands never got a rest. They had to interfere with the building-up of the bands and cut their supply lines. They had to provide a band-free area. Copying from Mao, the Germans declared fortified camps unsuitable targets for *Jagdkommando* activities, as also were strongly superior forces.

The *Jagdkommandos* were specially equipped for their tasks and, according to the regulations at least, well armed with many automatic pistols, automatic rifles, light machine guns, light grenade throwers and plenty of hand-grenades. They were supposed to fight and live for an extended period without additional supplies. 'Much

OPERATIONAL ASPECTS

cunning is required to lead a *Jagdkommando*', says the German Manual on *Warfare against Bands*. 'Exact knowledge of the fighting technique of the bands and of local conditions are a prerequisite for the successful application of crafty battle tricks. For this reason the band hunters should be deployed again and again in the district known to them' (No. 91). 'The *Jagdkommando* fights in the following manner: it marches mostly by night and moves into a hidden camp in day-time. March and rest must be protected. Reconnaissance begins when the battle area has been reached. From the footprints of the bands, the band activities are ascertained. In order to avoid treachery, no contact must be made with the population. Again and again the use of stationary reconnaissance troops has been found valuable for the *Jagdkommandos*. They observe the routes of approach and paths of the bands in places favourable for attack. Good camouflage, close liaison, and especially patience are the prerequisites of success. In order to maintain the element of surprise, the *Jagdkommando* does not transmit current reports. However, especially important reconnaissance results which call for immediate attack by stronger forces will be transmitted' (Nos. 92 and 93 of the Manual).

The *Jagdkommandos* had to march to their target area. In the Greek civil war, under British influence, commandos were created which specialized in lightning raids and quick withdrawals. The British had envisaged transforming them into airborne units, to be supplied from the air, but shortage of aircraft prevented the execution of this plan.

In Russia the Germans had resorted to another device, pseudo-gangs or counter-bands, composed of members of the German security police and the security service, police troops and reliable natives. They masqueraded as bands and their purpose was to check on the morale of the population and make contact with bands. In Palestine some pseudo-gangs worked on the British side during the post-war troubles, but their potentialities were fully realized only in Kenya, and there with outstanding effect.

It was a young British officer, Major Frank Kitson, who, assigned to intelligence out there, found that he had no intelligence to pass on and, unaware of precedents, thought of using pseudo-gangs, which he then proceeded to form. He introduced a startling new feature: no longer were they made up of his own nationals, with a few natives added to act as guides and screen, but they were gangs of captured ex-terrorists[1] with only one or two Europeans as commanders.[2] By imitating the enemy in appearance and behaviour these gangs could approach the enemy without arousing suspicion and fight at close range. They could obtain information and take immediate action. When this idea proved successful in the one or two districts where it had first been tried out, pseudo-gangs were formed and trained centrally, for use anywhere, and at this point Ian Henderson, a British police officer with unrivalled knowledge of Mau Mau ways, took over. As he put it, 'our technique of penetrating and living with other Mau Mau gangs proved immensely successful—time after time our collaborators contacted gangs and merged with them without difficulty'.[3] The pseudos ate and slept together with the real gangsters and then, at night, tied them up or kept them covered and led them into captivity. In this way twenty-two terrorists were accounted for each week. Most astonishing of all, 'the Mau Mau in the forests never had the remotest idea of what was going on. But it was not very long before the stage was reached when more than half the Mau Mau gangs on the Aberdares were actively working for us against their own leaders.'[4] Without resort to bribery and only by his understanding of Mau Mau mentality, Mr Henderson had converted many a captured bandit to his own use. What the Germans had half-heartedly tried in Yugoslavia, Greece and Russia—turning bands against bands—was perfected in Kenya to a fine art. It culminated in the successful hunt for the leader, Kimathi.

The Germans could have achieved the same outstanding result in Russia if their political leaders had not bungled matters. That the experiment could succeed in Kenya was due to characteristic traits in the Mau Mau make-up. 'For hours or even weeks', says Mr Henderson, ' a hardened supporter of Mau Mau will lean one way with utmost stubbornness, resisting every argument . . . then suddenly, some minute factor produces a fantastic change and the victim leans the other way, often with equal stubbornness.'[5] In other parts of the

[1] Similarly the French had used in Indo-China uniformed *Jagdkommandos*; ex Vietminh followers made up more than half the force. See also Bernard B. Fall, *op. cit.*, pp. 240 f.

[2] Cf. the most interesting account by Major Frank Kitson in his book *Gangs and Counter-Gangs*, London 1960.

[3] Cf. Mr Ian Henderson's fascinating book, *The Hunt for Kimathi*, London 1958. The reference here is to p. 163.

[4] *Ibid*, p. 168.

[5] *Ibid*, p. 261.

OPERATIONAL ASPECTS

world the anti-guerillas may have to fight their opponent by more prosaic means, such as the *Jagdkommandos*.

The other well-tried battle technique against guerillas is, of course, encirclement. Unlike *Jagdkommando* operations, it requires fairly large forces, since the lines must be strong, with advance protection and rear mobile reserves. In the words of the German Manual, 'the basic maxim of this technique is to cut off every escape route and then systematically to annihilate all parts of the band' (No. 73). The anti-guerillas assemble far away from the central band area and their departure must be so timed that they all reach the encirclement line at the same moment. The aim is to encircle the main band quickly and securely. While the encircling line is still in the process of formation heavy weapons must be moved to the front line so that the bands cannot break out at a weak point. The encircling line must at once be secured. Then follows the annihilation of the bands.

In Yugoslavia the Germans and Italians tried to encircle Tito's forces on several occasions. What the partisans have called the First Enemy Offensive was a German attack, at the end of 1941, against Tito's forces in Uzice in Serbia. The Germans advanced against the town from three sides, but it never came to an encirclement because Tito withdrew before the enemy columns had converged. In the course of the Second German Offensive, this time directed against partisan positions in east Bosnia, early in 1942, the partisans managed to break through the encirclement.[1] In May 1942 the Third Enemy Offensive was carried out by the Italians in Montenegro, but Tito again eluded the enemy's attempt at encirclement by withdrawing his forces.[2] The Fourth Offensive, a joint German-Italian operation, at the beginning of 1943, had as its objective the encirclement and annihilation of Tito's forces in Bosnia. Four German divisions attacked from the north and east, three Italian divisions from the west and south, while native units with Italian support tried to cut partisan escape routes to the south and south-east; but Tito, aware of the enemy's plans, cleverly deployed his divisions, forestalled the enemy and broke with his main force through the weakest enemy link, in the south-east, before the ring was properly closed.[3] When a second ring was closing on the partisans they made a feint of breaking out in

---
[1] Cf. Brigadier Sir Fitzroy Maclean, *Disputed Barricade*, London 1953, p. 169.
[2] *Ibid*, p. 185.
[3] *Ibid*, pp. 203, 208, and Vladimir Dedijer, *Tito Speaks*, p. 189.

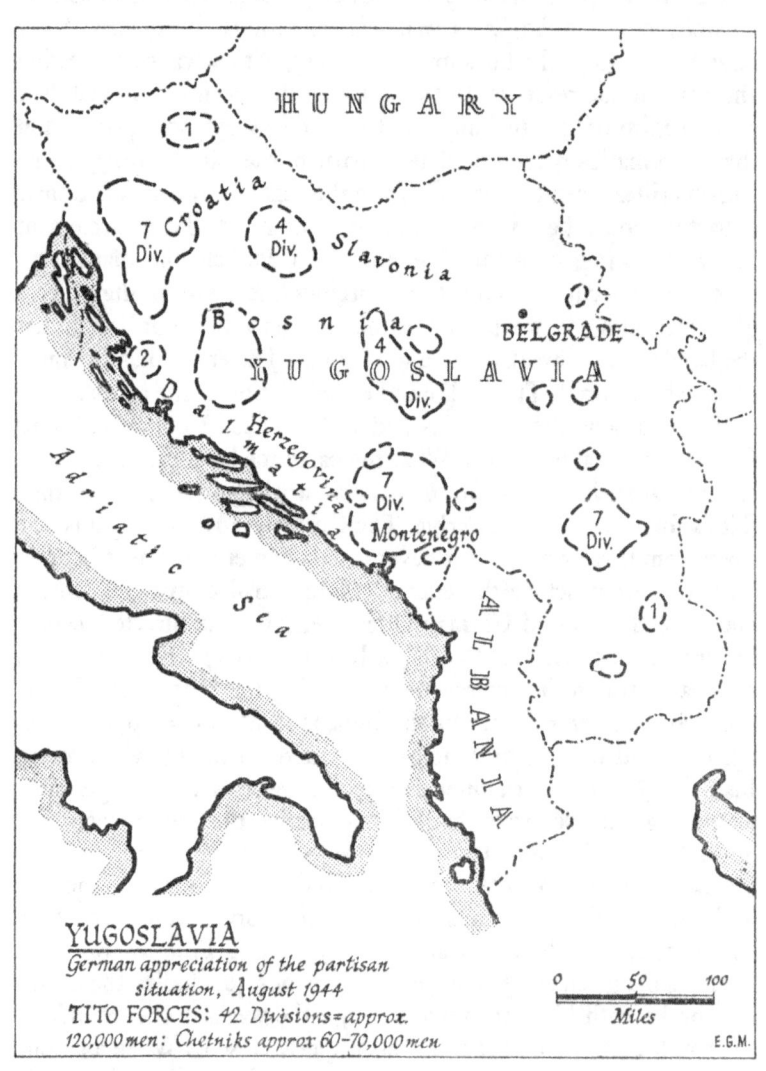

OPERATIONAL ASPECTS

the north, but actually broke through in the south where the opposing forces—Chetnik troops—were weakest.[1] In the Fifth Offensive in May-June 1943 in Montenegro, when more than 100,000 German, Italian and a few Bulgarian and native troops were fighting against less than 20,000 partisans, the preparations for encirclement were more thorough and the execution more ingenious.[2] The partisans found themselves completely surrounded but, leaving all heavy weapons behind, they ferociously attacked German units and fought their way out of the ring. During the Sixth Offensive, beginning in September 1943, the Germans rarely tried to encircle the enemy and the Seventh Offensive, in May 1944, was an attack on Tito's HQ by airborne troops and armoured and motorized forces.[3]

Partisan losses in some of these operations were very heavy—they lost 8,000 men in the Fifth Offensive—but the opponent never succeeded in annihilating the encircled forces. The partisans were either forewarned or received relief through diversionary attacks outside, managed to deceive the enemy about the point of their break-through attempt or directed it against the weakest link or fought more desperately. But especially in the Fourth and Fifth Offensive, their escape was very narrow.

We should therefore be wary of writing off encirclement and annihilation as failures in anti-partisan operations, even though only one of the ten major German actions against guerillas in Russia showed sizeable results, and British encirclement attempts in Malaya and those of the Japanese in China were not always fruitful either. In Korea, after all, encirclement operations against partisans succeeded.[4] It was found in Malaya, however, that not only was the large number of troops required for this operation impossible to control in badly mapped jungle, 'but also the build-up is so ponderous, the preparation so vast and lack of security so likely that, if any bandits have been stupid enough to remain in the area after "H hour" they will certainly have left it before the last "stop" and the last ambush party are in position prior to the advance of the striking force on to its objective'.[5]

[1] Brigadier Sir Fitzroy Maclean, *op. cit.*, p. 211.
[2] *Ibid*, p. 220; Vladimir Dedijer, *op. cit.*, p. 195.
[3] Brigadier Sir Fitzroy Maclean, *op. cit.*, pp. 249 and 257.
[4] Cf. Lt.-Colonel John E. Beebe, 'Beating the Guerillas', *Military Review*, vol. xxxv, December 1955, pp. 2 f.
[5] Major R. E. R. Robinson; 'Reflections of a Company Commander', in *The Army Quarterly*, vol. LXI, No. 1, October 1950, p. 80.

But it must be recognized that such a large-scale action as an encirclement should be attempted only if the band is large, and large bands will fight it out rather than disappear. Furthermore, it is clear that developments in aerial photography now make it possible to survey the battle area before the attack by means of reconnaissance aircraft which, flying at great height, no longer warn the bands, as hitherto, of an impending action against them.

An encirclement is a difficult operation to execute in any circumstances and it frequently miscarries in conventional warfare too. Nobody suggests that regular forces should not therefore attempt this operation against enemy regulars. Equally, there is no good reason why it should not in future be tried again against partisans. They did not, in the recent operations in Russia and Yugoslavia, use any tactics peculiar to guerillas in breaking out. It helped them that during the first operations the Germans kept too much to the roads. But what finally weighted the balance in their favour was that the partisans fought harder because they had more to lose.

But even if the partisans cannot be annihilated by encirclement or the employment of *Jagdkommandos*, they should still be attacked, if only to dislodge them and make them leave their present theatre of operations. The Germans in Russia considered such attacks not entirely useless because they relieved partisan pressure on communications for a short period until the band returned. But it should be recognized that if the band can be kept out of its old area and has to take up its activities elsewhere the advantages are more far-reaching, because it has been cut off from its channels of intelligence. Since all guerilla actions always depend for their success on good intelligence—for target selection, surprise, lightning blow, quick dispersal and retreat—a band removed from its old hunting grounds must reduce its activities or even remain inactive until it has again built up its intelligence service, and this requires time. This applies in particular to three of its intelligence sources, namely penetration, the network of agents, and monitoring. Between them, these are the mainstay of partisan information; through these sources the partisans are frequently briefed about possible objectives, and only then can most of the other intelligence sources be utilized. For instance, the agents inform the partisans in the first place about the intentions of the enemy, his change of plans, his intended dispositions, his future movements, his local sources of supply and so on, and it is frequently

OPERATIONAL ASPECTS

on the strength of this information that the other intelligence sources—reconnaissance, prisoners, captured documents—come into play. If a German corps report could state that the Soviet partisans were 'incredibly well informed of troop movements' and another report could say that the Soviet partisans learned of nearly every German troop movement early enough to plan an attack and set an ambush,[1] this was due to the three intelligence sources we have just mentioned.

Partisan intelligence, at which we propose to look now, serves two purposes: the partisans own and that of their regular army. For the army, the partisans carry out either a specific mission given to them by the unit on the opposite side of the front, or general tasks laid down in Field Service Regulations. The Field Service Regulations of the Red Army, 1944, designate as such tasks:

Observation and reconnaissance of the numerical strength of the enemy forces, disposition of his defensive installations and minefields;

Indicating targets for combat aircraft and long-range artillery;

Reporting to the command of the units of the Red Army the results of bombardments of important objectives ... by our aircraft; routes inaccessible or dangerous for tanks; places suitable for landing aircraft and for dropping landing parties (No. 861).

Furthermore, the partisans were enjoined to keep constant watch over the disposition and movement of troops and suppplies; determine the numerical strength of the enemy, the type of his weapons, the direction and time of his movements, the location of his troops and staffs, the names and numbers of his establishments and agencies; reconnoitre his aerodromes; determine the location, number and types of his aircraft, the aerodrome equipment, fuel and oil supplies, the security arrangements on the ground and in the air, the antiaircraft defences, the army depots and workshops; reconnoitre the defence lines, their armament, their organization and signal communications; and to capture orders, reports, operational maps and other enemy documents (No. 864).

Partisan intelligence for their own purposes has two objectives: to find out enemy intentions against them and to provide information

[1] Cf. Dixon and Heilbrunn, *op. cit.*, p. 93, and Oleg Anisimov, *The German Occupation in Northern Russia during World War II, Political and Administrative Aspects*. Research Program on the USSR, New York City 1954, p. 20.

PARTISAN WARFARE

for their own operations. They must also check the flow of information that reaches the enemy, and they do this by killing enemy informers, by terror methods designed to frighten the population away from the enemy, and by spreading misleading information about themselves.

The partisans obtain their information in the following ways:

1. Through penetration. Grivas had managed to penetrate the civil service in Cyprus,[1] while Soviet partisans had their own indigenous personnel at all levels in the German administration, from the policemen who guarded railway tracks and bridges and the officers of the indigenous security units which guarded other installations, to the village elders and mayors, who were often informed of anti-partisan operations and who in turn informed the partisans. There were even curious arrangements between partisans and German *Kommandanturas* whereby the partisans were notified of impending anti-partisan operations, while they in their turn kept the *Kommandantura* posted about movements of non-resident bands into their territory.[2]

2. Through their agents. These were in Soviet Russia either fulltimers or part-time agents who worked in their normal civilian employment as railway personnel or hospital staff, or in jobs with Germans as interpreters, secretaries, servants or so-called voluntary helpers in labour units.

3. Through reconnaissance by partisan detachments or partisan observers and interrogation of partisans returning from operations.

4. Through interrogation of prisoners and evaluation of captured documents.

5. Through monitoring enemy telephone conversations. The Soviet bands became past masters in this field and for this reason the German Manual on *Warfare against Bands* laid down special security precautions and discouraged the use of the telephone (No. 26).

The Soviet partisans took to monitoring only in the second half of the war; until then they had cut the wires. Monitoring certainly proved more profitable. It is unlikely that we have seen the last of

[1] Cf. Dudley Barker, *op. cit.*, p. 138.
[2] Cf. Office of the Chief of Military History, U.S. Department of the Army, *German Counter-Intelligence Activities in Occupied Russia* (1941–1944), n.d., p. 102.

this divergence from the original band activities of destruction, and in future the partisans, far from destroying enemy communications, might use them to confuse and misdirect him and disorganize his activities. A recent Swedish court case points in this direction.[1]

[1] London *Times*, June 19, 1952; O. Heilbrunn, *The Soviet Secret Services, op. cit.*, p. 86.

CHAPTER 5

GUERILLA AND ANTI-GUERILLA TACTICS

The various Soviet Russian regulations on partisan warfare used to be regarded as the best codification of guerilla tactics. Instructions for the commanders are contained in the Field Service Regulations of the Red Army, 1944, Chapter XVII, entitled 'Partisan Operations', while the *Handbook for Partisans, 1942*, the *Partisan Guidebook* and the *Handbook for Agents* give directives for all partisans. But recently the Viet-minh Manual on partisan warfare has become available, and it must now be considered the most authoritative and comprehensive statement on present-day guerilla tactics. While it reflects experience in revolutionary war, it is equally applicable to all other forms of guerilla warfare. In view of its unique importance and its practically unknown contents we need not apologize for quoting from it at length.[1]

The Manual lists as the seven principles of guerilla tactics intelligence, initiative, the will to attack, resoluteness, secrecy, speed, and perfection:

'What is required to fight intelligently? The following seven rules must be applied:

1. Make a feint attack on one point but actually attack somewhere else so that the enemy cannot protect himself. . . . For instance: harass the enemy clamorously in one place and fall upon him silently in another place. Give the appearance of having smaller forces than you actually have. Mislead the enemy, make him negligent, and then attack unexpectedly.

---

[1] The Viet-minh directives were first published by the Etat-Major de la Force Publique in Léopoldville in the *Bulletin Militaire* of June and August 1955 under the title 'Guérilla selon l'école communiste'. The *Bulletin Militaire* has rendered signal services to the students of guerilla warfare by its important publications in this field. The original text, from which the above translation has been made, is in French.

GUERILLA AND ANTI-GUERILLA TACTICS

2. Make yourself alternatively visible and invisible so that the enemy does not know where we are and cannot attack. For instance: come out of a secret tunnel in order to attack the enemy and then disappear at once. When the enemy enters a village, fight him for a time and then vanish. But when he retreats, pursue him resolutely.

3. Avoid the enemy's strong points and attack only his weak points; in other words, do not insist on confronting a numerically superior or watchful enemy with our entire forces. Catch him at his weak points: rearguard of marching troops, isolated soldiers who are resting or retreating. If we attack an enemy formation or an enemy on his guard, we are bound to suffer losses.

4. Know when to advance and when to retreat: if a stronger enemy attacks us violently, withdraw in order to try to counter-attack his weak points; for instance, wait until he is demoralized, tired or over-confident.

Never fight foolhardily or obstinately; do not become stereotyped or remain immobile in the same place. When your troops take up combat positions find out the routes of access and quick and easy retreat.

5. Attack, destroy and withdraw in such a way that the enemy cannot react, bring up reinforcements in order to encircle us or call upon his airforce. When the fight is over you must withdraw quickly and try to fight the enemy again and prevent his possible return. But for this speed we may be deprived of favourable opportunities, lose men and ammunition, and ourselves be encircled by the enemy.

6. Do not fight unless success is certain, otherwise withdraw. One must not attack foolishly, or persist regardless of cost, nor must one show useless obstinacy. If one wants to give battle one must be certain of success. If victory is not assured one must resolutely refuse to attack, and wait for a more favourable opportunity. If success is assured during battle one must resolutely assault in order to annihilate the enemy; but otherwise one must hold on for a while and then withdraw without regret or hesitation.

7. One must not always use the same tactics. The constant use of the same tactics allows the enemy to defend himself more easily and to gain the initiative. We must therefore vary our combat techniques, which requires intelligence and the constant recourse to ruses.'

Some of these dicta sound familiar; they were first laid down by

Mao Tse-tung. The Manual itself points out that it is based chiefly on the experience of the Chinese population and liberation army. The Manual, however, not only classifies and elaborates Mao's doctrines but adds many of its own and brings the subject matter up to date.

The second of the seven principles, initiative, implies that the guerillas must force the enemy to fall in with their intentions.

'In normal periods we must constantly seek to attack the enemy and not wait until he attacks us. By our diversionary and harassing attacks we exhaust him, compel him to withdraw and tie him down to his positions. If he tries to move out, we attack him at once and force him to turn back. If he tries a clearing operation, we either attempt to forestall him by attacking him or we slip away in order to avoid encirclement. In this way we can also ensure peace for the local people. During battle one must never stay immobile in one place, behind a bamboo fence or in a dugout. On the contrary, one must know how to move forward or back when necessary and to fight equally well inside and outside villages. If the enemy wants to move quickly, force him to slow down; if he wants to rest, force him to fight; if he wants to proceed on one road only, force him to use several so that he can be annihilated in small groups.

'The enemy must react against these techniques, and in doing so he must expose himself still more, thus offering us further opportunities for annihilating him. Equally, in every operation, one must try to envisage the enemy's reaction so that one can plan a new attack and make him comply with our intentions. On the other hand, one must consider whether the planned operation might endanger our bases and one must in this case make plans for their protection and the deployment of troops. If we tackle the enemy at his strong points or without precautions, we risk being thrown back and suffering damage. If we fight without thought for the protection of the population and our base, we risk losing both, and our ability to operate as well, because we are then without infrastructure.'

Here again the Manual elaborates a number of rules on initiative. They are:

'1. One must study the situation in the opposing camp, that is to say one must try to find out the dispositions of the enemy, the attitude of the cadres and the combatants towards one another, their morale and their fighting value. Follow closely the opponent's activities: the deployment of his intelligence service, transport, relief of pickets, all give evidence of his intentions and so make it possible to plan operations with every chance of success.

2. We must remedy our weak points: lack of arms and soldiers, fatigue, unfavourable climatic conditions. We must look after our arms; take them from the enemy, improve our obsolete equipment; ceaselessly spread propaganda in order to recruit partisans. Co-operate in combat with allied units and the guerillas of neighbouring regions; try to gain time so that you can await a favourable moment for action. Check the security arrangements and communications and liaison with the superior authorities, allied units and neighbouring villages. Keep fit. Seek favourable terrain. What matters most is strengthening one's fighting spirit, gaining knowledge, fusing all one's thoughts into the one desire to annihilate the enemy.

3. One must know how to discover the opponent's weak points; morale is low, the sentries are slack in the execution of their duties, there is general laxity, provisioning is difficult, duties are carried out mechanically, and so on. One must profit from these weaknesses in order to attack the enemy, and at the same time exhort the population to take an active part in the fight.

4. Achieve the impossible in order to compel the enemy to comply with our own intentions and prevent him guiding you. For example: if the enemy is about to attack us with large forces at a given point, we must evade him and then attack him in the rear. If he wants to enter the village without making his way through the gate, we must nevertheless force him to go through it and have sufficient forces ready there to wipe him out.

5. If the enemy protects himself carefully without presenting any weak points, we must create them before attacking. There are several possibilities:

'We resort to diversionary and harassing attacks in order to disturb and wear out the enemy; we encircle and split up his positions; we disrupt his lines of communication in order to make him come out to repair them, and then we attack him. In the course of battle we give the impression that we are forced to withdraw, so that the enemy takes up the pursuit and reaches our positions, where we wipe him out. We use stratagems and provocation to lead him on.

'In order to create weak points one must know the enemy's routine before deciding on the plan of attack. How are guard duties performed? How often does he receive provisions? Which route does he take to the village? How does he march or carry out movements and search operations? When one seeks to create weak points one must be secretive and persevering, not allow oneself to be led by the enemy and not give up plans simply because they are of secondary or only temporary importance. One must avoid the temptation to fight a big battle if one lacks the necessary means.'

The next section of the Manual deals with the will to attack, which it describes as the essence of initiative.

'If we allow the enemy to attack and restrict us to passive defence, we cannot protect the population and the troops are liable to be cut up. We must therefore endeavour to attack the enemy, force him on to the defensive, exhaust him, prevent him from increasing his radius of action, take the initiative from him and impede his endeavours. This is the way to achieve our glorious mission, "the protection of the population".'

The Manual gives the following directives for developing the spirit of aggressiveness:

'... We must apply the following precept: "If the enemy advances we withdraw; if he remains stationary we harass him; if he is tired we attack; if he retreats we pursue."

'If the enemy advances we withdraw: this must be understood to mean that if we are faced with a stronger and better prepared enemy we must temporarily withdraw to escape and counter-attack in due course. For instance, if we cannot hold out for a long time in our village in the course of a large-scale encirclement by the enemy, we must inflict losses on a part of his forces, withdraw into the neighbouring village or our underground shelters and then counter-attack the enemy forces when they are dispersed in order to rest or move back.

'On the other hand, if the advancing enemy lacks vigour and is not superior in numbers and matériel one must attack him and not withdraw.

GUERILLA AND ANTI-GUERILLA TACTICS

'If the enemy remains stationary we harass him: we give him no respite in order to tire him out and make him incapable of repelling an attack. The harassment does not always consist of firing to wound and kill a few, but is sometimes a real small-scale attack.

'We might also emerge from an underground hide-out to throw a few hand grenades into the village where the enemy is resting or attack the enemy on his arrival in a village when he is preparing quarters. If necessary, call upon neighbouring units for support.

'If the enemy is tired we attack. . . . We must now overcome our own tiredness and attack the enemy. Tiredness is the enemy's weakest point, because his morale is indifferent. One must make use of this opportunity to wipe him out. If we do not overcome our fatigue and so fail to attack, we miss a favourable opportunity and we may be attacked ourselves.

'When the enemy retreats we pursue: if the enemy takes flight after being knocked about by our troops, he must be resolutely pursued in order to wipe him out completely.'

The Manual has this to say on resoluteness:

'To be resolute means to be determined to attack the enemy swiftly whenever we can be certain of success. When the situation is unfavourable for us and necessitates withdrawal, we must resolutely withdraw in order to try and fight elsewhere. Any hesitation at critical moments can cause irreparable defeat.

'Resoluteness goes hand in hand with initiative. If, without danger to our bases and common plans, we can fight with a chance of success, we must do so without hesitation and without waiting for orders from higher authority. We might lose a good opportunity through the slightest hesitation.

'Normally, when one sees a weak point of the enemy which could be attacked, one must study the matter well before one starts the attack. On the other hand, if in the course of an operation the situation becomes unfavourable to us, we must immediately break off the fight. Courage is not synonymous with temerity or folly. If it appears in the course of an operation that the original plan needs modifying this must be done at once.'

The fifth tactical principle listed in the Manual is secrecy.

'Since enemy spies are everywhere, secrecy must be kept not only by our regular troops and guerillas but especially also by the population. The population must help to conceal us from the enemy's eyes and ears.

'Regular troops and guerillas must keep absolute secrecy before, during and after an operation, and about all their daily activities. Be careful about what you say and how you behave. Think before you talk and talk little. In the zone provisionally occupied it is advisable to lead the life of the local population, to move about and carefully to hide all things military. During troop movements one must avoid main roads, be careful in approaching assembly points in villages where one must reckon with barking dogs, and be absolutely quiet. When taking up combat positions speak little, move as little as possible, and do not moan when tired or wounded. Check before every troop movement that nothing has been left behind in the billets and that nothing reveals the change of position. Liaison and intelligence agents, when visiting the enemy-controlled zone, must not carry any documents except their instructions and the messages for transmission. Sentries must be on duty at all times. Precautions must be taken against traitors. Enemy spies must be annihilated. The cadres must set an example in keeping secrecy and they must also watch over their subordinates. The cadres and the troops must also educate the population to preserve secrecy, and familiarize it with counter-espionage.'

This is what the Manual has to say about the sixth principle, speed:

'In his rear, the enemy is in control of lines of communication, he has transport, phone and wireless contact, he can send reinforcements and can call on artillery and air support. This is his strong point.' The Manual then continues:

'In order to avoid the enemy's strong point and to transform it into a weak point we must be quick. Speed . . . gains us time and we can profit from favourable opportunities offered by the enemy to attack him successfully. One must be quick because the enemy will not remain negligent for long and, unless we exploit at once, we may suffer in the end (air-force and artillery support).'

GUERILLA AND ANTI-GUERILLA TACTICS

Then, after stating that arms, ammunition, mines, explosives, and so on must always be ready for action, the Manual goes on:

'During combat all movements, assault and retreat, and the firing of guns must be carried out quickly. Everybody must, through ardour and heroism, be equal to the enemy in morale from the very first minute of the engagement. In ambushes in particular the attacks must be made simultaneously in order to prevent the enemy from judging the situation correctly and to avoid missing a favourable opportunity. After each engagement the field of operation must be quickly cleared and matériel speedily recovered. In case of urgency the booty must be collected without delay. Take only arms and documents and withdraw at once.

'In order to carry out the operation quickly one must be thoroughly acquainted with the terrain, the roads and the nearby posts; one must be familiar with the plan of operation and one's task; one must know the combat technique thoroughly; finally, one must be instilled with the desire to annihilate the enemy and with belief in victory.'

What the Manual understands by the seventh principle, perfection, becomes clear from the following excerpts:

'The cadres and the guerillas must be resolved to destroy the enemy completely, that is to annihilate him and take all his arms from him.... It is of importance to determine before each operation the number of enemy soldiers to be annihilated and the number of arms to be taken, in accordance with our own means and the enemy's forces.... When we have the initiative in an engagement it is less advantageous to repulse a hundred or three hundred of the enemy than to kill two or three and take their arms.

'On the other hand, the complete annihilation of the enemy allows us to conserve our forces and to gain arms. The population will demonstrate its confidence even more strongly and the guerillas will profit from this. Also, we shall be able to fight a protracted war.'

Next, the Manual lays down that the essential element of guerilla tactics consists in attacking with the greatest violence and acting quickly. One must attack the enemy ceaselessly and everywhere.

'If the situation is confused, the enemy who cannot gain the initiative is reduced to passivity and at our mercy. Violence and speed of attack make it impossible for the enemy to defend himself properly and even more impossible to counter-attack. Through our heavy blows, quick movements and initiative the enemy will be pushed into fatal defensiveness and his forces will of necessity diminish while ours will grow daily.

'To attack continuously does not imply that one attacks day and night, without break. What it does mean is that as soon as an operation has been concluded one must think of the next one, prepare it carefully and execute it with the least possible delay. We must recognize this as our duty if we want to keep the enemy breathless and deny him the time to recover and reorganize. Co-operation on our part during combat with elements of neighbouring villages obliges the enemy to disperse his forces and allows us to act more rapidly. Continuous attacks and proper co-operation between elements of several villages or communities make it easier for us to discover the enemy's weak points. The knowledge of these weak points enables us to prepare in advance an entire programme for combined actions.'

After elaborating these points the Manual makes this observation:

'The enemy is superior to our guerillas in regard to technique, arms and numbers. In spite of this he often flees when we attack. He does not dare to advance unless he has air-force and artillery cover. This is due to the fact that his soldiers are not possessed with the will to fight; that they lack confidence in their forces and commanders; that they are not united among themselves.... We are inferior to the enemy in numbers. However, in spite of his bombs and the shells of his air-force and artillery we never hesitate to throw ourselves into the attack with complete disregard of death.

'In order to attack the enemy energetically and make quick decisions we must seek him out and attack him incessantly, always have the initiative, make use of all favourable opportunities in order to exterminate or weaken him. This is the essence of guerilla tactics.'

Finally the Manual stresses that the support of the people is a necessary ingredient of guerilla tactics. The guerillas have become increasingly stronger and better armed because of the support of the peasants, who realize that the guerillas are fighting to defend the life and property of the people.

GUERILLA AND ANTI-GUERILLA TACTICS

'Without the "popular antennae" (the French text uses the equivalent words) we would be without information; without the protection of the people we could neither keep our secrets nor execute quick movements; without the people the guerillas could neither attack the enemy nor replenish their forces and, in consequence, they could not accomplish their missions with ardour and speed. . . . As for the enemy, he cannot use guerilla tactics against us because he has no popular support and is fighting an unjust war. . . .

'The population helps us to fight the enemy by giving us information, suggesting ruses and plans, helping us to overcome difficulties due to lack of arms, and providing us with guides. It also supplies liaison agents, hides and protects us, assists our actions near posts, feeds us and looks after our wounded. . . . Co-operating with guerillas, it has participated in sabotage acts, in diversionary actions, in encircling the enemy, and in applying the scorched earth policy . . . On several occasions and in co-operation with guerillas, it has taken part in combat.'

So much for the seven principles of guerilla tactics. We now propose to consider what the Manual describes as 'lessons'.

One of the lessons is entitled 'Transmission and liaison'.

'The regional detachments and guerillas form only small units. Since, unlike [our] regular soldiers, they have neither wireless nor telephone communications they must use the population for transmission and liaison . . . It is advisable to create at village or community level a body specially charged with transmission and liaison; in this way secrets can be better kept and the peasants are not constantly disturbed in their daily work. In normal times as well as during periods of operation the cadres and the guerillas must take the occupations of the people into consideration when requesting them to do liaison work. They can ask the merchants, carriers and hawkers to hide documents in their packs in order to carry them to their addressees. They can use pre-arranged cries and words for sounding the alarm. One should communicate by signals, gestures or pre-arranged sentences. In order to apply this method successfully, villages and units in the same region must use a special system for alerts, questions and answers. The number of signals should be very limited so that the peasants do not forget them or mix them up. They must frequently be changed to assure secrecy.

According to conditions in a region we can organize a chain for transmission, we can select a confidential agent for the task of carrying statements of accounts or information reports to their destination . . ., we can use conventional signals such as drum beats, alarm bells, cymbals, rifle fire, bird calls, lighting a fire, hoisting a flag and so on, in order to keep friendly units informed about the position and the number of enemy soldiers, and we can install reserve liaison posts and chains. . . . All information must be passed on without delay to the superior authorities and to the neighbouring villages and units so that they can avoid the enemy and make their dispositions in time.'

Another lesson deals with troop movements and disposition of troops.

'The purpose of a move is to seek out the enemy for attack, to approach the chosen battlefield, to make propaganda in order to assist in the formation of bases, and to avoid for the time being a stronger enemy in order to conserve one's own forces.'

Then follow a number of detailed instructions about the necessary preparations for each troop movement and the protection of the column on the march.

'On arrival the troops must be allowed to rest, to study or prepare for battle, to spread propaganda among the population and to form bases. The commander must at once make contact with the regional committees, other guerilla detachments and friendly units, agree with them on measures for protection and defence and exchange information about the enemy. He should avoid returning repeatedly to the same place and making it a kind of permanent refuge. In particular, he has to watch over the morale of his troops, their health and food, and organize military and political studies for them. They must be in a constant state of combat readiness.'

The next lesson lays stress on the primary importance of continuous small actions which help the guerillas to keep the initiative, enervate the enemy, mislead him about the strength of the guerillas, make him negligent, and prepare the field for major operations.

Harassment is the subject of the next lesson. This comprises small surprise attacks, ambushes, propaganda activity among the enemy and annihilation of enemy forces. It requires neither large units nor good arms.

Next comes sabotage, which is directed against communications, telegraph posts, telephone wires, bridges, airplanes, ships, lorries, locomotives, guns and so on.

'Sabotage is a form of propaganda among the population, a stimulant in the fight against the enemy, and a duty of the population. Every inhabitant must consider sabotage as his mission and be determined to accomplish it. Sabotage must serve the interests of the people; that means its consequences must not unfavourably affect the life and production of the region. . . . Sabotage must not be undertaken if it could result in the destruction of the bases. One must instruct the population in every case how to encounter enemy reprisal actions and how to avoid fines or labour service.'

In order to carry out sabotage, the unit must be divided into two groups, one for destruction and the other for protection. The destruction group is again subdivided into several squads, each with a leader and a specific assignment. The Manual then continues with fundamental rules about the preparation and the conduct of this type of operation.

The next lesson, on mines, adds little to the knowledge of guerilla tactics.

The following lesson is on encirclement. It belongs in the category of tactical rather than operational planning. Unlike the encirclement carried out by anti-guerillas, it is only a small-scale action, and its military aspect is not the preponderant one; it has, in addition, politcial and economic aims.

'Encirclement aims at restricting enemy activities, patrols and search actions; intercepting food supplies, isolating the enemy and making life in the post intolerable. When one encircles a post one must also constantly harass the enemy and frequently organize a raid in order to take prisoner soldiers who leave the post singly. These activities throw the enemy into disorder, assist our propaganda amongst the enemy, and facilitate the clearing of the post. Encirclement gives us plenty of opportunities for surprise attack and ambush. It allows our guerilla units to increase their activities and to develop rapidly. It wins the local population over to our cause . . ., it prevents the people from selling foodstuffs to the enemy and assures the protection of the harvest, the intensification of production, and the consolidation and development of our bases. To sum up: encirclement requires a combination of the three forms of operation; political, economic, and military. If we encircle a post without inviting the people to fight in the political field, we cannot exercise our influence over the population, which will continue to work for the enemy, supply him with bamboo and sell him food, and our encirclement will not attain its aim. If we fight only in the political and military field without thought for the protection of the harvest, we expose the people to famine should the enemy retaliate.

'In order to isolate a post it is essential to obtain the co-operation of the people. Propaganda and meetings in which the enemy's crimes are denounced, will inflame the hate of the population and incite it to take part in the fight by refusing to work for the enemy or to deliver bamboo and food. The people's assistance is valuable to us from every viewpoint: information, propaganda, counter-espionage, spreading rumours. If our regulars and guerillas should undertake the encirclement operation all by themselves it would be ineffective, because with our tired and insufficient forces we would not be able to keep up our effort for long or to reassign the troops when necessary. During the blockade the regulars and guerillas must assist the cadres to form popular organizations, reorganize the bases, destroy renegade organs, suppress traitors and spies, instruct the inhabitants to hide their possessions and to keep the shelters against bombs and shells in good order, protect the harvest, encourage increased production and protect the population against enemy sweeps and reprisal operations.

'The forces required for an encirclement are not numerous. Moreover it is necessary regularly to relieve the forces in order to avoid fatigue. Our forces must be split up in small units and an observer group must be formed whose task it will be to offer resistance to enemy raiding parties, to intercept reinforcements from other posts and to capture enemy informers. First we deploy our forces far away from the post, then we approach slowly. The nearer we come to the post the more difficult it is for the enemy to fire. Each group must seek protection in dugouts and frequently change its position. The groups protect each other if the enemy attacks. Our actions must never become repetitive . . . and the form of our activities must also be varied: at times we harrass, then we carry out a *coup de main* or hold a propaganda meeting among the enemy. . . . During the encirclement we must be well informed about the situation and judge it correctly in order to come to the right decision; we appraise the situation with a view to tightening or widening the encirclement ring; the main point is to know how to protect the population against losses and devastation; on the other hand, the widening of the ring must not adversely affect our popular bases.'

The Manual then lays down the following rules:

'(a) *If we tighten the ring:* If we tighten the ring again our aims are to mount a surprise attack and wipe out the post, to ambush reinforcements and supply convoys, to intensify our propaganda among the enemy whose morale is already affected and finally to annihilate those who run away. During this period we multiply our harassing and commando operations. We capture isolated enemies. We harass the post in various ways and without respite. We organize rifle detachments against aircraft in order to deny the enemy food supplies dropped by parachute; we sabotage communication lines; we lay mines to prevent the enemy from leaving the post; we spread misleading rumours; we use various propaganda devices among the enemy; we arrange for letters from their relatives to reach partisan rebels; if possible we invite relatives to appear before the post and appeal to the rebels to rejoin the ranks of the resistance. While we are tightening the encirclement we must avoid the more "brutal" actions which could rouse and provoke the enemy who, to avenge himself, may terrorize the inhabitants, with unfortunate results for our popular bases and the intensification of production.

'(b) *If we widen the ring:* Our aim in this case is to draw the enemy, to induce him to leave the post so that we can exterminate him, and to protect our bases if he chooses to offer determined resistance.

'When the order for widening the ring is given the population must spread false information so that the enemy who is always influenced will seriously believe that our troops are withdrawing and will therefore move out of the post. We arrange an ambush to assault the enemy.

'As long as the enemy still has strong forces and offers determined resistance to our assaults, we must take the decision to widen the ring in order to assure the protection of our bases. We must also order the widening of the ring during the harvest: we must save the population from being terrorized by the enemy, who will not hesitate to impede production.

'We must reassure the people by way of explanation when we decide to widen the ring. We must then try to camouflage our bases. We must have plans for protecting the people against enemy reprisals. We must intensify our propaganda among the rebel soldiers. We must take advantage of every opportunity to create a cell in the enemy ranks. In brief, we must prepare to tighten the ring again when we have the chance.

'In the course of an encirclement operation we must know the enemy's situation and arrange for close contacts with the villages on the one hand and the different echelons on the other, so that we can efficiently co-ordinate our efforts and keep each other informed.

'We are then in a position to seize a favourable opportunity for launching an all-out attack against the enemy's position, for ambushing enemy troops who intervene or withdraw, or for deciding to break off the encirclement in order better to defend our bases.'

The Manual explains in its next lesson ambush tactics against enemy infantry. It describes the ambush as the principal battle technique for regional troops and guerillas. This technique allows small forces to attack strong enemy forces and to exploit the element of surprise by attacking the enemy with lightning speed and annihilating him. 'It is characteristic of our determination always to keep the initiative and to launch unimportant but successful attacks.'

'The ambush has several advantages: it causes us only small losses, yet allows us to capture arms and replenish the weapons of our unit. It therefore recommends itself as a very convenient technique for newly formed units with insufficient arms.

'It is essential to form and develop popular bases if one wants to ambush successfully. Thanks to the inhabitants, we are well informed about the enemy situation and we can secretly deploy our forces and thus surprise the enemy.'

Secrecy, surprise and speed are, in the words of the Manual, the three ingredients of success. 'Since the ambushing technique exploits the element of surprise, it is necessary to keep absolute secrecy in the planning stage as well as during the execution. The assault force must be so deployed that it can encircle and split up the enemy. Fire must be concentrated on one main point in order to increase the momentum of our assault. Furthermore, the assault must be carried out quickly. One must attack with speed, "get on with the job" quickly, so that the enemy does not know how to defend himself. Surprise and speed in execution keep our losses to a minimum and reduce the effectiveness of enemy aircraft, artillery and tanks.'

The Manual differentiates between two types of ambushes, the one in which the guerillas take up their positions in the terrain as they find it, the other where the ground is first prepared by 'terriers' who dig holes and dugouts and camouflage them carefully prior to action; it all depends on whether the ground is suitable or has first to be made suitable for the ambushing operation.[1]

The actual procedure is the same in both cases. The Manual is most insistent on careful preparation. The leaders in charge of the operation first seek information from their agents about the enemy situation and about their own popular bases. Then, accompanied by their staffs and some participants in the planned action, they inspect the terrain. If they decide in favour of the operation, plans must be prepared; these plans must cover the smallest detail: aim of the operation, difficulties, advantages, possible developments during the action, instructions on how to overcome the difficulties; the plans must also determine the required arms and ammunition, calculate the number of shovels and pickaxes as well as other matériel required for crossing water-courses and so on. Then follows the spiritual preparation, 'a very important aspect'.

At the assembly point the troops are briefed in every detail about their mission. The troops then march to their assigned stations, if possible using different routes and avoiding populated districts. Some troops are detailed to deny the road to the head of the enemy column, others to cut off the retreat of the enemy's rear; others are assigned to harass the enemy, and others again to carry out the assault. Since the aim of the operation is to split up the enemy forces into several parts and wipe them out, the assault party must be the strongest. The commander is stationed near the fire party and he is in touch, through signals or agents, with the observer post which re-

[1] Before Dien Bien Phu the Viet-minh dug many trenches from their hills to the French fortress where they branched off to either side to give width for the attack.

ports the approach of the enemy and possible enemy reinforcements.
Fire is opened when the enemy arrives.

'The enemy, suddenly and brutally attacked, becomes disorganized. We rush, encircle and split up the enemy. We prevent him from regrouping or dispersing and fleeing. . . . That is, if the operation goes according to plan. . . . Otherwise, if our troops have already taken up favourable positions, if the terrain is unfavourable for the enemy, if our cadres and fighters keep up their morale we continue the engagement, decide to assault in force, and we will be successful. We no longer intend completely to annihilate the enemy as planned; we try instead to destroy those enemy forces which have penetrated into our positions, while we harass and contain his other troops.

'But if the enemy is too strong and our success is not assured we withdraw without hesitation, or hide and await his return (for then he will be over-confident and likely to underestimate our forces) . . .'

Finally: the withdrawal.

'Once success is achieved we must be quick: we assemble the deserters and prisoners, we collect the booty, we destroy what we do not need, we evacuate the dead and wounded and then order the withdrawal. We leave in several groups and use different routes in order to avoid the enemy air-force and artillery and make quicker progress. The withdrawal must be organized; it must not become a kind of "sauve qui peut". One group is detailed to contain the enemy and assure the withdrawal of our main force to the pre-arranged assembly point.'

The next lesson deals with surprise attack and commando raid. Both are directed against enemy forts or stationary troops.

'If we have rather large forces available, sufficient completely to wipe out the garrison of a post or a security force stationed in a fixed place, we call our offensive a surprise attack. But if we employ only a few troops or a half-section in order to attack a weakly guarded objective or a few enemy soldiers we call it a "commando raid". While "harassing actions", big or small, are meant to exhaust the enemy, his manpower, his ammunition and so on, and to undermine his morale, surprise attacks and commando raids have as their aim his complete annihilation.'

Both these operations rely for their success on the assistance of the fifth column.

'As a rule the morale of the enemy partisans and of his militia is very low. It is easy for us to find "intermediaries" and to create cells among them. . . . In this way success is assured and, moreover, our losses are very small. Indeed, once the enemy soldiers and rebel partisans are convinced by our propaganda and won over to our cause, they become our helpers. They skilfully co-operate with us; while our fighters assault the outer defences, they sow the seeds of discord inside. . . . It is therefore necessary for our soldiers and guerillas to give special attention to propaganda among the rebel soldiers. We must induce the population to agitate among the rebel soldiers, partisans, militia and so on. We must above all take advantage of the time when the enemy suffers losses and his morale is shaken.'

Full use is to be made of stratagems.

'We must act above all when the enemy, full of self-confidence, is underestimating us. We then order our men to disguise themselves as coolies, as enemy soldiers, as hawkers on their way to the market place. Our disguised fighters must exploit the element of surprise in order to wipe out the enemy in his fort or garrison. This technique requires constant and detailed information; one must be fully informed about the enemy's situation, from the first preparations to the time for execution.'

Even if the guerillas, for some reason, have to attack suddenly and without preparation, propaganda still plays its part. 'Do not forget to use intensive propaganda methods during the attack in order to sap the will to resist.' But in the main the guerillas have to rely in such a case on fire power.

The Manual then elaborates six conditions necessary for success. The first is full information about the enemy troops, their arms, morale, fighting spirit, their daily activities, routine, guards, patrols, possible reinforcements, the terrain, and the morale of the population. Next comes a minutely prepared plan; then speed and secrecy; the selection of the principal target for the main force which, if conditions are very favourable, may encircle the enemy; quick, unobserved and co-ordinated approach; and finally quick withdrawal. The commandos, in particular, go into action in mufti and mix with the population afterwards to escape the enemy's attention.

After a victorious engagement comes the formation and consolidation of popular bases.

'Since the aim of the surprise attack is to wipe out the enemy, eliminate traitors, destroy renegade organizations and win the population over to the cause of resistance, our plans must provide for the formation of bases after the attack. If our existing bases in the region are relatively satisfactory we leave behind after a successful attack a small cadre which must co-operate with the population and the regional cadres in order to suppress spies and informers. We must take preventive measures against terrorism (by the enemy) and thwart his designs for re-establishing his reactionary militia. However, if we have no bases or if they are weak, we cannot leave any of our friends behind; our soldiers as well as our propaganda cadres must go back during the following nights in order to carry out propaganda activities among the population. The population will be asked to form itself into popular organizations; popular group detachments and guerilla units must be created. In short, the inhabitants are progressively led into the fight against the enemy, a fight which must be undertaken in all forms.'

The final lesson deals with the dispersal and concentration of forces; these are treated as a means of keeping the intiative.

'If the enemy troop dispositions disclose weakly guarded points and the opportunity is favourable to us, we must concentrate our forces in order to launch an attack. On the other hand, if we need to divide our forces, whether in order to create bases or to escape the enemy's sweep, we must without hesitation give the order to disperse. Whether we are concentrated or dispersed, we must have a plan which foresees all possibilities and lays down our intentions. A dispersed unit must be able to concentrate quickly and vice versa. Only by alternatively dispersing and concentrating can we always keep the initiative.

GUERILLA AND ANTI-GUERILLA TACTICS

'The purpose of dispersal is to avoid superior enemy forces and to conserve one's own forces, to co-operate with guerillas and popular troops in order to carry out isolated attacks, and to form popular bases.... But while the troops are dispersed they must maintain close liaison with the main force as well as with other dispersed groups. Thanks to this contact the dispersed elements can easily regroup when a favourable opportunity arises.

'On the other hand, concentration is required in preparation for an attack, for a joint mission with regional and regular battalions, and "when the situation permits and for short periods only, in order to study, exchange information about one's experiences, or simply for rest and recreation". The degree and duration of concentration depend on circumstances. "Generally speaking, it is inadvisable to keep strong forces concentrated for too long a period, because we risk becoming bored and inactive. If the enemy is stronger the period of concentration must be as short as possible. If, for example, we want to attack, we do not order the concentration until the plans have been prepared and well studied. Once the combat mission has been executed we quickly review the conduct of the operation and then instruct our forces to disperse without delay."'

The lesson concludes with an interpretation of the meaning of personal initiative. 'One must develop personal initiative in each section, which must not rely exclusively on superior orders.... Let us add that the term "personal initiative" requires elucidation. Only in special cases, when superior directives cannot get through in time, are subordinate elements authorized to take appropriate action on their own initiative. They must in all other cases ask for directives and execute the orders of the superior echelon; they must never act in accordance with their own ideas.... Some people do not like the concentration and prefer dispersal. According to them, disperal allows them to disappear easily when the enemy begins a sweep. But there are others who think only of concentration. They are worried by dispersal because they are afraid of isolated combat. If the enemy clears the district they must dig themselves in in trenches and dugouts, and they are afraid of this too. Furthermore, in case of dispersal, they might be sent to a remote corner of the enemy-occupied zone where they would have to endure much privation and hardship.

'This erroneous tendency must be corrected (that is the tendency to act according to one's own preferences in case of dispersal). This so-called freedom of action runs counter to the spirit in which superior orders have to be executed and to the respect for discipline.'

So much for the Manual. It is repetitive at times, but then, it is addressed to simple people and aims at being clearly understood. If it discourages initiative, it does so for the sake of the reliability of the movement as a whole. That the lessons were learnt is beyond doubt. The Viet-minh propaganda among the population and enemy troops was successful, their intelligence service in particular was first rate, and their operations were well executed.

Partisans operating in more developed countries will of course engage to a much larger extent in sabotage actions, particularly against roads and railways, telephone and telegraph lines, transport, depots, power plants, industrial enterprises and other objects of military importance. They will also lay ambushes for railway trains and attack enemy held villages. The Soviet partisan instructions already mentioned contain very specific directives about all these types of operation. The technique is by now sufficiently well known and does not require recapitulation here.[1]

The tactical aims of the guerillas are threefold. In the first place they want to draw enemy forces, in national wars in order to weaken his front line, in revolutionary wars in order to destroy him piecemeal, impede his task of pacification, and prevent him from efficiently protecting the population and winning it over. In the second place, they intend to weaken his infrastructure, and the third task is to win the support of the people.

The essence of their tactics is to select weak objectives for their attack, and to offer the enemy no target for his counterstroke.

Several well-known factors make these tactics possible. The enemy has to protect the population and guard his bases and communications. Part of his forces are therefore stationary and, as it were, sitting targets; others, regularly moving supplies by road and rail, are in almost the same position. The guerillas, however, protect and guard almost nothing and possibly the only time they used the railway

---

[1] For the Field Service Regulations cf. O. Heilbrunn, *The Soviet Secret Services*, *op. cit.*, pp. 188-191, and for the Handbook for Partisans see O. Heilbrunn, *Partisanenbuch*, *op. cit.*, pp. 60-63.

for their own transport was when after the German raid on Tito's HQ in a cave, he and his staff boarded a train and managed to escape the German pursuit.[1] The guerillas, if attacked, just vanish; in a national war they move to another camp, in a revolutionary war they move to another base. The enemy, who has to split up his forces and employ many of them on guard and convoy duties, finds it difficult to collect superior forces for attack, while the guerillas can more easily do so. Again, wearing no uniform and being natives of the country, they can easily disappear among the people, while the enemy soldier will always be conspicuous, as will the enemy's modern paraphernalia of war. It all adds up to the same thing: the enemy offers targets and a number of them are bound to be vulnerable, yet if he wants to attack, the guerillas are just not there.

A regular army would be in a better position to fight guerillas if it could adopt guerilla status and guerilla techniques, but this it cannot do. If it is fighting in a national war against a well equipped enemy it too must have modern weapons and therefore be dependent on bases and communications. If it is fighting in a revolutionary war it cannot but protect the population and again for this task, as Colonel Woodhouse has pointed out, it needs good communications. Even if a regular army relied on air transport it could not entirely solve the problem; many aerodromes would be necessary for this purpose, requiring large numbers of guards. A regular army will always offer many targets and those manning them will always be recognized as soldiers whether they wear uniform or not.

But that does not mean that a regular army cannot or should not improve its ways, and that goes for defence as well as attack.

In national wars the regular troops must hold a line. In the revolutionary war in Indo-China the French Union forces also tried to hold a line, or rather a number of lines and, having insufficient forces to man them properly, they tried to make do with forts as substitutes, with negative effect.

The moral is obviously this: if a revolutionary war is fought by determined guerillas and the anti-guerillas have insufficient forces, they will help no one but the enemy if they try to hold a line. The anti-guerillas must, in this case, get away from the conception that all wars are fought of necessity on a long, continuous front.

What the regular army needs instead are bases, in the same way as

[1] Mao's semi-regulars, however, frequently used the railways.

the guerillas have bases. Dien Bien Phu would have fulfilled the purpose of a base, had it been densely populated and better located.[1] The bases must be so located that they can ensure the protection of a large section of the population and if necessary the inhabitants of outlying regions must be evacuated to these bases. They must be well prepared against guerilla attacks and be secured by heavy weapons. They must have a well protected airfield—if need be, indigenous labour must be excluded—and aircraft for observation, liaison, evacuation, transport and fire support. The base must be secured by *Jagdkommandos* roaming around at a fair distance from the base[2] and, if it can be arranged, by pseudo-bands. To burden the *Jagdkommandos* with heavy arms would be a mistake; it has rightly been asserted that the Task Force 100, operating on the French side in Indo-China and composed of motorized infantry, an artillery battalion, a tank squadron and a service unit, was slowly wiped out because it had too much of the good things and could never keep pace with the enemy marching on foot through the jungle.[3] If the base is about to be encircled and cannot hope to be relieved or to resist, the main garrison must break out of the ring in guerilla fashion and keep moving, as Colonel Rigg has suggested.[4]

Like a guerilla base, anti-guerilla bases must be consolidated and expanded. Guerillas have shown that they can gain a country in this way and there is no reason to suppose that the anti-guerillas could not do likewise.

The number of bases to be formed must be related to the forces available and the attitude of the population. It depends on this attitude how far the population itself will fight the guerillas and raise reliable self-defence units in the communities. The aim must always be to shift as much of the defence burden as possible on to the population.

Highly mobile reserves, possibly airborne, are stationed in the bases for quick intervention anywhere. They must be large enough to allow of constant deployment of part of their personnel for en-

[1] Dien Bien Phu was situated on low ground and surrounded by hills which were not incorporated into the defence system.

[2] These suggestions have been made by Major BEM L. Marlière, 'Quelques leçons à tirer des Campagnes du Laos', *Bulletin Militaire* 1955, p. 561.

[3] As Professor Bernard Fall has pointed out in 'Das Ende der Kampfgruppe 100', *Wehrwissenschaftliche Rundschau* 1960, p. 611.

[4] Cf. above Chapter 4.

GUERILLA AND ANTI-GUERILLA TACTICS

circlement, *Jagdkommando*, pseudo-gang and fighting patrol operations.

Finally, stationary troops must be kept to a minimum and all available forces mobilized for attacking guerillas. Posts should be installed only along the lines of communication and for their protection, and the German system in Soviet Russia might serve as an example. Larger support units for quick assistance in case of an attack on a post are stationed at suitable intervals.

The anti-guerillas have therefore three missions which they must fulfil simultaneously. They must defend the bases and posts, and they must also defend installations which are vital to the anti-guerillas or must be denied to the guerillas, such as land and sea communications with a foreign country supporting them. They must attack the guerillas, in particular their bases. And finally, they must win the support of the population.

For the general deployment of the anti-guerilla forces the following scheme, devised by the Ecole d'Application de l'Infanterie of Saint-Maixent,[1] deserves attention. It is based on the degree of guerilla activity in the various parts of the country.

'In zones where there is little trouble, which are well under control and where communications are secure, posts of all kinds are numerous but of minor importance; their area of control is large. Police and security forces will suffice for manning these posts. Small forces are adequate, support forces are centralized.

'In rebel areas the posts (bases) are strong and less numerous; they are not in control of any territory. Communications are difficult and dangerous. This is the zone where strong forces, band reconnaissance and widespread offensive actions are required.

'In intermediate zones the posts (bases) are rather numerous and strong; they must control a larger territory. Their support is a hard task. Reserves must be strong. This is the most active combat zone. Here one must keep the population and the rebel communications under close surveillance.'

The choice of organizational structure for the anti-guerillas depends in the first place on whether they are fighting on foreign soil or in their own territory. There are again several possibilities in the first case. During World War II anti-partisan warfare in Yugoslavia,

[1] Reprinted in *Bulletin Militaire*, December 1957, p. 855 f. The above translation comes from pp. 866/867.

Greece and Albania was entirely the affair of the German military, while in Russia the army was responsible for anti-partisan warfare only in the combat zone and the army group rear areas; Himmler was put in charge of all anti-partisan operations in the *Reichskommissariats*. If the anti-partisans are fighting in their own territory they may either appoint the C-in-C to the post of Commissioner General, as the French did in Indo-China, or they may make the military director of operations responsible to the Governor, as for instance the British did in Cyprus.[1] The important point in these cases is that the military, police and civil authorities must work together as a team, and this close co-operation was assured in Cyprus through the military director's Combined Staff, which consisted of representatives of all three departments, and the subordinate district security committees which again consisted of representatives of the military, police and civil affairs.

As far as the defensive aspect of anti-guerilla warfare is concerned, it is obvious that the troops themselves must be protected and, naturally, they have to provide the protection themselves. Before troops move through guerilla-infested territory, the commander must familiarize himself with the guerilla situation, and the German commanders in the various World War II partisan areas did this by consulting the band-situation maps. These maps were provided at the so-called information collecting centres of the local commanders responsible for anti-guerilla warfare. The band-situation maps, compiled by the local intelligence officer, gave as much information about the guerillas as HQ maps about the opposing enemy troops. The march itself must be protected in the usual way: scouts go ahead into the villages on the march, a mine detection troop marches in front of the column, and automatic weapons, kept at the ready, must be distributed over the entire length of the column. Billets and camps must be protected in all directions. According to the German directives, the troops must be so distributed that everybody can reach the scene of action by the shortest route. Troops must not first fall in at alarm posts because this procedure allows the guerillas to open fire on the concentrated troops and it would only delay the defence.

The German Manual for Anti-Band Warfare gives very detailed

[1] For the set-up in Malaya see 'NOLL, The Emergency in Malaya', *The Army Quarterly*, April 1954, p. 59, and for Kenya General Erskine 'Kenya—Mau Mau' in *Royal United Service Institution Journal* 1956, pp. 11 f.

GUERILLA AND ANTI-GUERILLA TACTICS

instructions for the protection of posts but the principles are well known and do not require description. The techniques for the protection of the lines of communication, installations, etc., have also become a familiar subject.

As for offensive operations against partisans, we have already discussed encirclement and the employment of pseudo-bands, *Jagdkommandos* and ferret forces.[1] If neither sufficiently large forces for encirclement operations nor specialized forces in the form of *Jagdkommandos* are available, guerilla units are destroyed by surprise attack and hunt. The aim of this technique is, by surprise attack, to compel the guerillas to give battle, to split them up quickly and to hunt and destroy the split-up groups. As the German Manual adds laconically: 'One has to put up with the escape of individual isolated groups.' It recommends this technique as especially successful against bands which are not yet familiar with the place or which march through the area.

The leading idea behind this technique is borrowed from the guerillas: superiority in the decisive place makes it possible quickly to split up the opponent and to pursue him until he is wiped out. And again, prior reconnaissance is imperative: before the assault starts the positions and strength of the opponent must be ascertained.

In order to achieve surprise the troops assemble far away from the scene of action and advance quickly on roads already unobtrusively reconnoitred. If it can be assumed that the guerillas will avoid a fight, the marching column has to go over to the attack at once; only if the guerillas will in all probability offer serious resistance should the troops first take up positions for the attack. If the guerillas accept battle, concentrated fire is directed against the guerilla positions, they are stormed and broken up and the defeated enemy is wiped out. But if the guerillas avoid battle, the troops follow only slowly in line behind the retreating enemy, but move up the flanks quickly so as to encircle him. If the guerillas split up, the troops take up the hunt. The various battle groups must be correspondingly strong. They must unrelentingly keep up the pursuit of the guerilla groups and

[1] While encirclement and the large-scale employment of pseudo-gangs are techniques of an operational nature, the purist can rightly point out that *Jagdkommandos* and ferret forces are tactical striking forces and should therefore have been treated in this Chapter. Since they are usually employed together with other means such as large-scale resettlement actions, their operations, by this combination, are of operational significance, and it was therefore more convenient to treat them in the previous Chapter.

compel them to fight. If the guerillas disperse, the operation must be stopped and a fresh reconnaissance made. It can be assumed that the guerilla leader has instructed his detachment to reassemble later far away from the scene of action. It is a reconnaissance task quickly to find the assembly points. A new plan for attack must then be made.

The hunt is described in the Manual as a more intense form of pursuit. Its aim is to overtake the guerillas, make them fight it out and annihilate them. The troops must be highly mobile and it is therefore recommended that ammunition and supplies be brought up by other troops. The hunt is above all directed against the guerilla command and, if possible, the guerilla leader must be taken prisoner or killed.

There is another form of attack which was used by British forces—the so-called egg-beater. Here a large number of patrols stirs up the partisans in the hope that they will stumble into one of the patrols. Finally there is the patrol sweep, also of British origin, where again a large number of patrols are deployed but their main task is to detect signs of enemy presence and of his movements.[1]

All these operations require relatively large forces. Lieutenant-Colonel Clutterbuck has emphasized the fact that small-scale operations—ambushes and patrols—have on the whole been more successful in Malaya, Kenya and Cyprus, because they exploit existing intelligence or the troops procure it themselves.[2] This is undoubtedly correct. It must be remembered, though, that in these countries partisan concentrations were never large, and what we said about anti-guerilla encirclement operations in Chapter 4 is applicable here too: large-scale actions are, as a rule, useful only if the partisan forces are large. There appears to be only one exception and this is the cordon and search, an operation which of necessity requires many troops and cannot successfully be carried out piecemeal by a few soldiers.

The German Manual also contains special directives for the deployment of the air-force in anti-guerilla warfare, a subject with which we shall deal in Chapter 8. Little use has so far been made of armoured trains in anti-guerilla warfare and it therefore seems appropriate to draw here on German experience in World War II.

According to the Manual, armoured trains are specially useful for

[1] Cf. Lt.-Colonel R. L. Clutterbuck, 'Bertrand Stewart Prize Essay 1960', *The Army Quarterly*, January 1961, p. 165.   [2] *Ibid*, p. 166.

GUERILLA AND ANTI-GUERILLA TACTICS

the fight against guerillas. In large-scale operations armoured trains are called upon to penetrate into partisan areas and carry out independent operations, generally to supply artillery support, to transport reconnaissance and other tanks, mortars or support units, to prevent the escape of partisans over the railway tracks, and in exceptional cases, when the armoured trains cannot be actively deployed, they can be used as command posts. Armoured trains are especially effective if used on their missions by reconnaissance, security and scouting patrols, since they keep the partisans in a continuous state of alarm. Armoured trains can also be used for supplying posts along the track, transporting the wounded, protecting very vulnerable railway stations, bridges, etc., protecting repair squads and transporting troops.

The Germans also employed individually-powered single armoured units on railway tracks. Ten to fourteen of these units, travelling at a distance of about 500 to 600 yards from each other, formed a special armoured 'train' and were used for bringing quick support to endangered places on and near the line.

Bearing in mind that guerilla territory in Soviet Russia was often near or around railway lines and that these were their main target, the effectiveness of armoured trains in this kind of anti-guerilla warfare cannot be doubted.

The Manual is rather reticent about the employment of tanks in anti-partisan operations; it says that their value lies not only in their fire power but especially also in the fact that they deeply impress the opponent. It must be added that the anti-guerillas had here a weapon which the guerillas had so far usually been unable to employ themselves. The Germans used tanks particularly during the fifth, sixth, and seventh offensives against Tito in Yugoslavia, in what may be called their conventional role of highly mobile artillery. Other countries used tanks in other theatres of anti-guerilla war on convoy and road patrol duties. Owing to the difficulties of the terrain in which guerillas usually operate, the impact of armoured columns on the conduct of anti-guerilla operations seems likely of necessity to remain limited.

Guerilla war has so far been a foot-soldiers' war, and the most important anti-guerilla requisite is the capacity to march on foot at least as fast as the opponent. Moreover, the anti-guerillas need many such foot-soldiers to succeed.

The third tactical task of the anti-guerillas, to win the support of the population, is so closely connected with the question of how to treat the population that we shall discuss it, together with this subject, in Chapter 10.

CHAPTER 6

GUERILLA AND ANTI-GUERILLA TECHNIQUES

Some guerilla and anti-guerilla techniques have already been touched upon in preceding chapters, together with operational or tactical aspects, and we will now deal with the remaining items, especially guerilla attacks on railways, installations and convoys, and with anti-guerilla sweeps.

1. Railways. The simplest form of attack against a railway is by way of sabotage, and the saboteur's choice is wide. Sabotage may be used against the track by mining the rails or by knocking off their screw heads; the rolling stock can be sabotaged by introducing sand into the grease-boxes of carriages or explosives into the fire-boxes of engines or by removing the fish plates, then bending one rail to the inside and wedging a fish plate between the two rails; and if the railway is electrified one can demolish the pylons with explosives, break the insulators with a shot, damage the turbines or smash the instrument board and the roof insulators of electric locomotives. A veritable textbook on partisan sabotage has recently been published and we refer the interested reader to this source for further information.[1]

Railway tracks are the favourite guerilla target for ambushes, preferably in deep cuts, on slopes, curves, and high embankments; other objects, such as bridges, signals, water hydrants and reservoirs or railway stations, are usually better guarded. If a train is ambushed, telephone and telegraph wires must be cut before its arrival, and if possible the scene of action should be far away from stations or other locations where the enemy has reinforcements. The main group of the detachment takes up its position opposite to where the middle of the wrecked train will stop. Another group with machine-guns and

---

[1] Cf. Hauptmann H. von Dach Bern, *Der Totale Widerstand, Kleinkriegsanleitung für jedermann*, Schriftenreihe des Schweizerischen Unteroffiziersverbandes, Biel 1958, pp. 51 f.

sub-machine guns places itself at the tail of the train on both sides of the track so that it can conveniently fire along the railway cars. If enough forces and time are available, two reserve groups take up positions some hundred yards above and below the scene of the incident in order to stop enemy reinforcements. Strict fire control is necessary to prevent the guerillas hitting each other.

For attacks on railway and other installations the detachment is usually divided into three groups—an assault group, which removes the guards and destroys communications, a demolition group which destroys the target, and a reserve group. The first object of attack is the communication centre, which must be seized and destroyed. The Viet-minh, however, operated without the reserve group.

If a railway station is to be attacked, the detachment is again divided into two or three groups. The principal target of the demolition group is usually the signal box, the telephone communications and the tracks, especially switches and frogs. Particularly if the objectives are close together, the sequence of action must be well timed. The first object of attack is again the telephone and telegraph system.

2. An enemy convoy is ambushed in a place which offers good cover to the guerillas and makes it difficult for the enemy to deploy his vehicles, especially tanks, to the side of the road. The first and the last vehicle are destroyed simultaneously, with mines, hand-grenades or machine gun fire. Then fire is opened point blank and distributed against the entire column.

If the enemy pursues the ambushing force it withdraws to a previously specified point of assembly, which must be located some way from the base.

The ambush technique used against enemy infantry has already been described in the previous chapter.

3. The Soviet Partisan Handbook contains special directives for attacks on villages. The technique is more or less the same as that used in conventional fighting. Other regulations in various manuals on sabotage, ambushes and raids and the techniques actually used by guerillas everywhere have become too well-known to require further description.

Turning now to anti-guerilla techniques, we shall discuss, for the same reason, only the encirclement technique or rather techniques. The Germans in fact developed four techniques in World War II

GUERILLA AND ANTI-GUERILLA TECHNIQUES

for annihilating encircled guerilla detachments; we will describe these methods briefly.

The simplest one is called battue shooting. All encircling forces advance simultaneously towards the centre. This technique can, however, only be used in small areas; a long encirclement line, required for larger areas, can hardly be expected to advance equally fast and the partisans could easily escape through the resulting gaps.

The second technique is called the 'partridge drive' in the German Manual. In the words of the Manual, 'the forces of one sector of the encircling line advance, while the forces opposite remain in their position. The attacking forces press the guerillas, as in a partridge drive, towards the stationary forces. One must reckon with the possibility of the guerillas trying to escape through the attacking force. In order to prevent this, reserves must be deployed at a sufficient distance behind the attacking forces. This method is recommended if the guerillas' direction of escape or their escape routes are known, and also where a part of the encircling forces has taken up position in a particularly easily defensible area (river, plateau, forest fringe), so that attempts to break out in this place are bound to fail and would quickly result in the guerillas' annihilation' (No. 78(*b*) of the Manual).

The third technique attempts to drive in wedges immediately after the encircling troops have taken up position. Strong forces of mixed units advance from the encircling line towards the centre of the ring or the known guerilla camp while the other encircling troops stay put. This technique does not give the guerillas a chance to discover any weak parts in the encirclement ring and, if they are energetically attacked, they are compelled to split into several groups. At this point the cordon advances whilst moving forward strong flanking forces, takes up contact with the wedges and splits up the ring into several smaller rings which are cleared without haste. Should the fighting go on for several days, the troops must be given daily objectives. These must be reached before dusk. Far-advanced wedges must organize all-round protection and so must strong wedges, while weak wedges return for the night to the main forces in the line (Manual, No. 78 (*c*)).

Finally, the fourth technique calls for the formation of a shock unit, and this technique is applicable if the guerillas have a permanent camp which they presumably want to defend. After the encirclement ring has been formed, the troops hitherto used as reserves form a strong assault group and, advancing from the encircling line, attack the camp and wipe out the guerillas. The forces in the ring stop fleeing guerillas and sweep the area later on.

The German encirclement techniques have in the meanwhile been developed by the use of aircraft, as will be seen in Chapter 8.

CHAPTER 7

RELATIONS WITH THE REGULAR ARMY

Before we discuss the relationship between army and partisans, we ought to decide whether partisan forces are really necessary. The answer appears obvious, both for the East, which has exploited this kind of warfare with great success, and for the West, which has trained or supplied and directed partisans in many parts of the world and profited from their exploits in the last war. Yet partisan tasks can be entrusted to more regular units, and the Norwegians did this when they decided to form a Home Guard.

The Norwegian Home Guard is not intended to replace the regular army, but to supplement it by carrying out certain defence duties, security and reconnaissance tasks, guard and scout duties, and, in addition, it shall 'prevent or delay enemy transport of men and supplies by operations behind the enemy lines, in the flanks or by attacking his transport and supply convoys', and offer 'armed resistance in occupied territory in the form of guerilla activities and sabotage'.[1]

The Home Guard can undoubtedly do what partisans can do, and since large numbers of personnel have been specially trained for guerilla activities the Home Guard can probably carry out these tasks more efficiently. Yet, being uniformed, it will find it more difficult than guerillas to melt into the countryside. Furthermore, the Home Guard is a part of the Norwegian military defence system, its Inspector-General is a member of HQ Norwegian Forces, and the Home Guard units are under the command of the three Services. If a regular army is defeated in the field and has to lay down its arms, the victor will insist in an armistice agreement that all parts of the military defence system, including the Home Guard, be disbanded and possibly interned. A Home Guard cannot therefore replace

[1] Cf. *A Survey of the Norwegian Home Guard*, issued by the Inspector-General of the Norwegian Home Guard, Oslo, May 1955, pp. 12 and 13.

guerilla forces of the type which, on the European Continent and elsewhere, took up the fight in the last war after their regular army had surrendered.

The relations between the army and a partisan movement may be distant, close, or anything in between, and in their determination geography plays a decisive part. The underground movement in Poland, almost beyond range of Western aircraft, was too far away from the theatres of operations of its friends to receive much support in its heroic struggle. Again, geography may be the tactical factor which decides the timing and degree of support. As long as the fighting in Africa lasted, the German supply lines went through the Balkan countries, and when Italy was invaded it became the Balkans' task to pin down the German occupation forces which would otherwise have gone to Italy. Maximum assistance and close liaison, through British and American Military Missions, was therefore decided on. Elsewhere it may be desirable to postpone large-scale partisan activity and massive support until the invasion is about to take place—as happened in the case of France, Belgium and Holland. Finally the availability of means, the military or political appreciation of the partisan situation, the partisan contribution to the war effort, and elements of chance may also determine the relationship.

The practical forms of co-operation, apart from sending supplies, may be summarized as follows:

1. Liaison officers and agents may be sent to allied partisans; we have already referred to such cases.

2. Army officers may train guerillas. This was done by Britain in Burma in 1941 in the Bush Warfare School, and in 1943 British officers were parachuted into Malaya for the same purpose.

3. Army officers may take command of detachments, as British officers did in Malaya, Borneo and Italy, and American officers in the Philippines.

4. Army officers may form or take command of the entire movement, as General Bor did in Poland and General Mihailovitch in Yugoslavia. Or an official Military Action Committee may steer the movement, as was the case with the French and Belgian Resistance.

5. An army unit may co-operate with partisans:

(a) *In special assignments.* Before the outbreak of the Korean war North Korean guerillas stationed in the South suddenly became active in order to denude the frontier of South Korean troops; and when

RELATIONS WITH THE REGULAR ARMY

they were successful, the North Korean army attacked and broke through the thinly held lines of its opponent.[1]

(b) *In battle*. In August 1940, in the 'Hundred Regiments Battle', the largest combined force ever assembled in guerilla war, numbering 500,000 Chinese, smashed the Japanese communication lines throughout Northern China.[2]

(c) *In campaigns*. In Wingate's Ethiopian campaign in 1941 indigenous partisans supported the regulars throughout and finally, reinforced by some regulars, blocked the Italian escape route before Agibar, while Wingate's forces pressed them from the other side against the partisans and obtained the enemy's surrender.[3] On a much larger scale, in the Soviet summer offensive of 1944, the partisans had the task of slowing down the German retreat after the Soviets had reached the Dnieper, and split up the German lines so that the Germans would have to withdraw on the roads and railways and could be blocked by partisan forces there until the Red Army had moved up and could annihilate them.[4]

Intelligence co-operation is possible in two ways:

(a) The partisans collect and pass on information to the army. This practice was adopted in much of occupied Europe during the last war.

(b) The army seconds its own intelligence officers to the partisans, as the Red Army did from 1943.

There are numerous other examples of army-partisan co-operation; the SAS and the Special Boat Squadron, the US 'Burma 101' and the Australian Independent Companies all worked with partisans, and the semi-regular forces in China and Indo-China constantly co-operated with them. The above examples will suffice as illustrations.

Wherever there are partisans in the field there must be co-operation with the army, for two reasons. In the first place, without such co-operation partisan action may be harmful to the army: the partisans may destroy the bridge which the army wants to find intact in

---

[1] See for the Korean guerilla fighting Lt.-Col. John E. Beebe, 'Beating the Guerilla,' *Military Review*, vol. xxxv, December 1955, pp. 2 f.

[2] Gene Z. Hanrahan, 'The Chinese Red Army and Guerilla Warfare', in *U.S. Army Combat Forces Journal*, February 1951, p. 12.

[3] Cf. W. E. D. Allen, *Gideon Force*, in Irwin R. Blacker, *Irregulars, Partisans, Guerillas*, New York 1954, p. 385.

[4] Cf. Major Edgar M. Howell, *op. cit.*, p. 196.

its advance, and tear up the roads and railway tracks on which the army may wish to follow quickly the retreating enemy. The partisans may engage in large-scale operations and attract enemy reinforcements to an area where the army wants to break through; they may prevent enemy reserves from moving to the place on which the army is making a feint attack, and so forth. In the second place, partisan forces everywhere, co-operating with the army, have achieved a certain measure of success and helped to lighten the task of the regulars.[1]

Two conditions must be fulfilled, however, to make this co-operation acceptable to the army. First, the sacrifices which partisan war demands from the partisans as well as the population must be justifiable and the partisans must therefore submit to control. Notable examples of such conformity were the French and Belgian resistance movements. It was realized that any premature partisan activity on a large scale in these countries would do more harm to the population than good to the war effort. French resistance activities were therefore divided into two phases, the Immediate Action phase and the National Rising phase. Immediate action comprised sabotage and the fight against enemy agents, undertaken by groups of a few people only, and these small-scale operations were carried out prior to the invasion. The National Rising included raids against transports, enemy columns and command posts, large-scale sabotage and prevention of enemy sabotage, and coincided with the Allied landing in Normandy. Similarly, the Second Section of the Belgian General Staff in London had made plans for a two-stage partisan war. The country was divided into five zones and the order was given that intensive sabotage should start when the Allies were within thirty miles of the zone and large guerilla operations should begin when the Allies entered the zone.[2] It is to the credit of these movements that on the whole the orders were followed, whereas on the other hand Mihailovitch failed, in the Allied view, to take sufficiently forceful action, and in Greece British Liaison Officers found it difficult to

---

[1] Cf. *The Memoirs of Field-Marshal Kesselring*, London 1953, p. 231: 'But they (the Italian partisans) were only a vital menace where they were directly co-ordinated with military operations.'

[2] For the organization of the French Resistance see Colonel Demange, 'La Guérilla', *Revue Militaire Générale*, February 1960, pp. 224/5, and for Belgium Jean Temmerman, 'La Résistance Belge', in *L'Armée, La Nation*, January 1955, p. 33.

co-ordinate partisan efforts with the general strategy of the Middle East High Command.[1]

Secondly, while irregularity is a partisan characteristic, it must not be carried too far. This is a question of the quality of the leadership and of the rank and file, and of the cohesion of the movement. If the inception of partisan movements is left to chance, if in particular they are not prepared before the emergency or outbreak, such deficiencies must be accepted because one cannot expect self-appointed leaders and untested men to form good teams.

There are, however, a number of reasons for preparing a partisan movement beforehand. In the first place, suitable leaders can be appointed and rivalries eliminated. Secondly, one may not have the time to form a movement afterwards. It took the Soviets eighteen months to find an appropriate structure for their partisan movement; a future war might long be over by then. Also, it may not be possible to organize the movement afterwards because all official and unofficial agencies which can carry out this task, may have ceased to function. The Soviets had the underground party organization and the NKVD to do the job; the West would have no comparable institutions to draw on. Furthermore, even if one has the time and the means to organize the movement after the outbreak, the partisans cannot make any contribution to the war effort in the form of operations and intelligence during the possibly decisive first few weeks of the war. An enemy will also find it easier to suppress the movement if it has not even been activated when he invades the country. And last but not least, unless the movement is prepared beforehand it may easily be infiltrated by an enemy's sympathizers and fellow-travellers since their trustworthiness can no longer be checked in occupied parts of the country.

As against this it has been said, especially in Switzerland, that one would give the show away if the movement were prepared beforehand—warn the enemy of what was afoot, enable him to liquidate the movement and seize its arms, equipment and supplies before it had fired a shot. If this argument means that the enemy would know for certain that he would encounter partisan resistance, it has not much force behind it, because every invader will nowadays take this possibility into account. It may also mean that large-scale partisan re-

[1] Cf. Colonel C. M. Woodhouse, *Apple of Discord*, London 1948, and Brigadier E. C. W. Myers, *Greek Entanglement*, London 1955.

cruiting in peace time would enable the enemy to obtain detailed information about the organization through indiscretion or treason. But a partisan will not know more about the partisan movement than a regular soldier knows about the army, and this argument therefore does not carry much weight. On balance the case for advance preparation is strong.

Advance preparation requires in the first place a decision on the type of co-operation envisaged. For the dispatch of liaison officers to friendly partisans abroad no preparation is needed. If they are to be guided and supplied, the SOE and OSS organizations have to be revived. If a country intends to have its own partisan movement, at home or in its overseas possessions, a survey is necessary in order to establish which parts are suitable for partisan war. It can then go all lengths, but minimum preparation demands that the leadership be made familiar with guerilla tactics and methods.

This implies that the structure of the partisan movement and its relation to the army have been decided on, that the decision has been made whether to appoint officers or partisans or both to staff and command positions, that at least the higher hierarchy has been appointed and preferably that a Field Manual for Guerilla Warfare has been prepared.

Decisions must also be made at this stage about arms, equipment and supplies. Requirements must be assessed and the question settled what and how much should be issued beforehand to partisan storage depots. Camp sites must be selected and depots built.

To recruit and train the rank and file in peace-time will in practice be possible only if a framework such as the Norwegian Home Guard is adopted for this purpose. But it is of the greatest importance that a sufficient number of trained radio and wireless operators be available. Finally, a skeleton security organization, especially for screening partisan recruits, must be set up.

For the West only two schemes for army-partisan co-operation are feasible: the partisans are either subordinated to the army, or they both work in co-ordination under a Defence Committee. The Soviets in the last war chose the latter structure, possibly because they did not want the Red Army to become too powerful, or because they wanted the party to share in the glory, or because the partisans—a medley of Communist fanatics, Red Army soldiers, workers and peasants who sought escape from Soviet or German terror by joining

RELATIONS WITH THE REGULAR ARMY

the movement, and genuine volunteers—could be better handled by the Party and NKVD than by the Red Army.

The problem of whether to subordinate the partisans to, or co-ordinate them with, the army was first discussed by General Denning.[1] In his words

'guerillas used for our own purposes behind the enemy's lines must operate under the orders of the Commander-in-Chief in the Field. If this is not done, their action will not contribute towards the main object, which is the defeat of the enemy's forces in the field; in fact it may impede the attainment of this object. Control and direction of guerillas by a governmental ministry would be going back to the days of "private armies" which, however gallant and determined they may have been, did not make the contribution which they might have made, had they been under the Commander-in-Chief in the Field. It is only the Commander-in-Chief who can decide the guerilla action which will help him to attain his object.'

This seems indeed to be the right solution, for two reasons. The enemy troops fight against our own troops and partisans under *one* commander, their C-in-C. To put our troops and partisans under different commanders would only simplify the enemy's task. But above all, we must recognize that the battlefield is no longer a five-mile front as it was at Waterloo, or several hundred miles of front as it was in World Wars I and II, that is to say, a combat zone which has been extended in length; modern partisan warfare has extended it also in depth, from the enemy's front line to his base. This is the extent of the battlefield in total war, where the enemy is attacked at the front, the flanks *and* in the rear. It is *one* battlefield, and there can only be one C-in-C on one battlefield.[2]

This was also the system adopted in China. In the large expanses of the country it was not everywhere feasible to integrate the guerilla detachments into the army, but wherever possible, and particularly in the base areas, the military took charge of the guerilla units in the

---

[1] In Foreword to Dixon/Heilbrunn, *Communist Guerilla Warfare, op. cit.*, p p. vi. and vii.

[2] It appears from various publications by former Soviet partisan leaders that the R d Army frequently gave direct orders to the partisans. Cf. P. K. Ignatow, *Partisanen*, Berlin 1953, pp. 159, 177 and 423; A. Fedorov, *L'Obkom clandestin au travail*, vol. ii, Paris 1951, p. 320; S. A. Kovpak, *Our Partisan Course*, London 1947, p. 58.

PARTISAN WARFARE

area.[1] Similarly in Indo-China the Viet-minh commander of the regulars was, for important operations, in charge of all troops, the regulars, the regional battalions and the guerillas.[2] It cannot be doubted that this system worked well in both countries.

A Table of Organization, which provides for the subordination of the partisan movement to the army, appears in the first chapter of this book as Chart 1.

As for the structure of the partisan movement, there can be no question of grouping the formations into divisions, corps and armies. Formations of this type are only suitable for independent guerilla movements and not for those fighting as auxiliaries to the army. The largest Soviet partisan formation was called a brigade. The strength of these brigades varied: the 3rd Leningrad Brigade had 2,000 men, the 4th only 400. But there were also, at least until the end of 1943, a number of partisan detachments which were not grouped together into brigades, and in the northern sector one such detachment had only 150 men, while another was supposed to be 1,000 strong.[3] A brigade had from two to six battalions and a battalion two to four companies. Brigade Staff had an operations section, a supply, intelligence and signal section and others.

It will probably depend on the length of the front line of partisan units stationed immediately behind the enemy's front whether they are to be subordinated to army brigade or division, while partisan units further in the rear will probably be subordinate to the officer in command of the army rear area. No useful purpose could be served by going into further details.

Should officers or civilians be appointed to partisan staff and command positions?

The task of a successful partisan leader is a difficult one. He must recruit and train his men. Training must be extensive because the partisans must be able to do what the infantry, the engineers, the pioneers, the signal troops, supply and medical units can do. The partisans must be able to use not only their own arms but also those of the enemy. The leader must procure arms and food, and he must arrange for reserve camps and depots. He must keep liaison with the

[1] Gene Z. Hanrahan, *The Chinese Red Army and Guerilla Warfare*, loc. cit., p. 12.
[2] Colonel de Grevecoeur, *La guerre du Viet-minh*, Tropique, June 1953, p. 12.
[3] *Feldzug gegen die Sowjetunion der Heeresgruppe Nord, Kriegsjahr 1943*. Oberkommando der Heeresgruppe Nord, December 24, 1944.

RELATIONS WITH THE REGULAR ARMY

army and neighbouring units. He must know the enemy's intentions, select his own targets, provide his reconnaissance, and attack. He must serve as army intelligence officer unless the army seconds its own personnel. He must recruit intelligence agents. He must keep up the morale of his unit and maintain discipline.

The general trend in countries where partisans have fought with the army seems to have been in favour of appointing army officers, and this was the case even in Soviet Russia, which ought to have preferred party nominees. At the beginning of partisan resistance probably most of the leaders were communists, either party functionaries or people who occupied leading positions in civilian life. From early in 1942 Red Army officers were flown in to reorganize partisan detachments and also to command and staff larger units; the latter increased in numbers as the war progressed. In China the detachments were led by army officers specially trained in partisan warfare, while the North Koreans relied for guerilla warfare to some extent on the army itself; army elements that had been by-passed, strengthened by an infiltrated division, were supposed to carry on as partisans. Only the Viet-minh do not seem to have appointed semi-regulars to partisan commands.

There are several reasons why on the whole officers seem more suitable than partisans to command and staff these forces. With the exception of former partisan leaders, they are the only class whose individual suitability can be judged on the strength of service records. They can better fill staff appointments because this is part of their métier, they know the army ways better, and they can prepare themselves better in peace-time than outsiders. If a regular army has to lay down its arms trained officers can better absorb the army remnants into the partisan movement and enforce discipline than any civilian could. And if ever, after the elimination of their army, the partisans have to fight on alone, officers must lead them. But above all, since the partisans take their orders from the army, their command should conform to that of the army.

For army intelligence military peronnel should be attached to the partisans. While civilian members of the Resistance, working in small groups, did magnificent intelligence work, partisan detachments were less reliable in this field. There was a tendency to inflate enemy numbers in their reports or to announce that areas were cleared of the enemy which in fact were not. Military personnel will be free from any temptation to magnify partisan achievements. No less important, they are familar with army operational requirements.

It may be said that partisan forces, even under army control and army officered and staffed, will never reach the standard of Special Forces and that they absorb a great number of officers who would be more profitably employed in the Special Forces.

During the last war Special Forces frequently used guerilla tactics and methods against guerilla targets.

'Swooping swiftly and silently out of the desert they (the Special Air Service) carried out their blitz raids and then, just as suddenly, vanished again into the uncharted regions south of the coast road. Axis convoys, driving along unsuspectingly in the safety of their own back areas, suddenly found themselves under withering fire from the machine guns of SAS jeeps. Planes, dispersed on their aerodromes, burst into flames as SAS time bombs exploded. In all, more than 400 planes were destroyed by these adventurous raiding parties.'[1] The SAS, in their magnificent exploits, cut roads and blew up railways behind the enemy lines, they laid ambushes, cut telephone wires, destroyed lorries and railway trucks, blew up ammunition dumps and created diversions, they reconnoitred for the army and reported many targets to the RAF. The Long Range Desert Group attacked lines of communication, forts, petrol dumps and convoys, and Peniakoff's force was engaged in similar exploits. The Chindits cut road, rail and river communications, assaulted enemy installations and prevented the passage of enemy reinforcements. On the opposing side the Brandenburg Division was first deployed, as a battalion, in sabotage tasks behind the Polish lines and later on in the Caucasus, near Lake Aral and in the area between the Ural and the Volga. Almost all the Special Forces had learned to fire enemy weapons and they had become experts in handling explosives.

They had, of course, other tasks outside the partisan sphere of activity, but between them they have handled every task partisans can do and they did most of them better. No partisan detachment ever equalled them in dash, discipline and determination. Yet Special Forces cannot replace the partisans. The Special Forces operated mostly in territory where there were no partisans; in the desert, in Italy—prior to the armistice—in Germany and in Burma where the

[1] D. I. Harrison, *These Men are Dangerous*, London 1957, p. xii.

RELATIONS WITH THE REGULAR ARMY

'Independence Army' was won for the Allied cause only in 1944. Also the Brandenburg Division was only engaged in partisan activities where the Germans had no partisan support.[1] When British Special Forces fought in partisan areas, on the Dalmatian Islands, in France and Italy, they linked themselves with the partisans, although the link in France was rather loose. No army can spare from the front the many soldiers which the Special Forces would require to attack the enemy behind the lines in depth. While Special Forces can do guerilla work in a limited number of places at a time, the guerillas know no limit: twenty-four hours before the Red Army attacked in June 1944 in the Brobruisk-Vitebsk area, Soviet partisans carried out 10,000 raids and thoroughly destroyed all German lines of communication.

Where there are no partisans or where the population is hostile, Special Forces will frequently be indispensable. If Special Forces are required in partisan territory their work will be complementary; the Special Forces might stiffen partisan resistance and do the precision work, but the partisans are needed for mass output, and while Special Forces can only spend a limited time on any particular job the partisans, being stationed in the area, can keep up the work indefinitely and mine a road or railway track for weeks and months on end. Since they are, as a rule, more numerous than Special Forces and a permanent menace to the enemy, they will bind very much larger forces than the Special Forces can do, and will also tie down the enemy permanently. China, Soviet Russia, Yugoslavia, and Indo-China are the most obvious examples. Hence while partisans may compete with the Special Forces for officers, and outstandingly good ones at that, they certainly deserve them.

When Special Forces and partisans seem equally well qualified for the execution of a particular mission, it should be kept in mind that the latter usually have a number of advantages on their side. They are stationed behind the enemy front, they therefore need not cross the enemy's lines and return through them when they have completed their mission. Since they are nearer to the target they are less likely to be discovered by the enemy during their approach march, they can more easily reconnoitre before the attack and therefore gain

---

[1] In some of their operations they tried to organize partisan resistance. For 'Operation Schamyl' where they went behind the Soviet lines to the Caucasus for this purpose cf. Major Herbert Kriegsheim, *op. cit.*, pp. 90 f.

a clearer idea of how best to attack. They know the terrain well and can make better use of it. On espionage missions too most of the advantages are with the partisans. When questioned by the enemy they can more easily justify their presence, they can more easily get the identification papers prescribed by the enemy, they speak the local language, and they have their agents in the enemy's offices, messes and billets. Finally, the supply problems of partisan forces are usually easier to solve than those of Special Forces.

If one considers how often well-led partisan forces have defeated regular armies or caused them serious trouble, then the advantages of having well-led partisans on one's own side seem rather obvious.

CHAPTER 8

THE AIR FORCE IN GUERILLA WARFARE

In the last war there were not many occasions on which the air force supported partisans in combat. Most of the help was given to Tito's partisans, and when the German attacked them, in May 1944, the Allied air forces quickly provided direct air support, flying in a single week more than 1,000 sorties against targets connected with the enemy attack.[1] Apart from taking 11,000 wounded partisans out of Yugoslavia, Allied air forces subsequently struck mainly against rail traffic, and when the partisans were again attacked, in July 1944, in Montenegro, the air force supported the counter-attack by an onslaught on German troop concentrations.[2] It then switched its attention to Albania where close support was given to the partisans, and by the beginning of 1945 it was back again in strength over Yugoslavia, attacking targets on the battlefield, troop concentrations and communications. The air force then supported the newly formed Fourth National Liberation Army in the final operations for the liberation of Dalmatia and Istria,[3] attacking strongpoints, gun positions, headquarters, troop concentrations and so on. For these operations air advisers were attached to Army HQ and RAF liaison officers were assigned to each corps. The officially commissioned RAF History considers the part played by the Allied air forces in these operations as decisive.[4]

On the other side of the world Philipino guerillas received considerable help from the American Air Force early in 1945.[5] In order

---

[1] Hilary St George Saunders, *Royal Air Force, 1939–1945*, vol. iii, *The Fight is Won*, London 1954, p. 236.
[2] *Ibid*, p. 238.
[3] Colonel Ilija Zelenika, *The Yugoslav Airforce*, Air Power, Spring 1954, p. 279.
[4] Hilary St George Saunders, *op. cit.*, p. 250.
[5] Cf. for the following Joe G. Taylor, 'Air Support for Guerillas on Cebu', *Military Affairs*, vol. xxiii, No. 3, 1959, pp. 149 f.

to achieve close co-operation, Philipino officers were attached to Air Force HQ and five American air support teams were sent to the guerilla detachments. These air support teams, equipped with radio sets mounted on jeeps, could request air support; they also acted as ground control to direct air attacks.

Close support operations in the Philippines were not always successful, because the terrain made such operations difficult, but on one occasion at least success was achieved when a Japanese column, advancing towards an airfield held by the guerillas, was driven back by machine gun fire from the air; the guerillas were able to exploit the advantage and recaptured a position lost in the fighting. For the rest the American Air Force carried out bombing and strafing missions: guerilla scouts would follow Japanese troops on the move and pass on to the AAF the location of their bivouacs for the day, which were then bombed. In the guerilla attack on Cebu City the air support team on top of a mountain had telephone communications with observers in the guerilla front line who gave targets to the AAF—troop concentrations, supply and ammunition dumps, shipyards, submarine pens etc.—all indicated by guerilla intelligence.

On the Soviet front only one incident was reported in which the Red Air Force co-operated with partisans in an engagement, and this was Operation *Frühlingsfest* (spring festival). In this operation, almost the last German action against partisans, Stuka formations assisted the German attack, while the partisans received from their air force tactical support and glider-borne reinforcements. After heavy fighting the partisans, in spite of their stubborn resistance, were forced to withdraw. In their counter-attack they suffered heavy losses. They tried to break out but only a few thousand succeeded. Their casualties are believed to have numbered 14,000.[1] Air force intervention does not seem to have brought the partisans any substantial relief on this occasion, and it is not known whether there was any exchange of information between ground and air during the operation.

The air forces made their main contribution to the partisan war in another field, that of supply missions. In 1940 Special Operations Executive was formed in Britain with the object of giving help to the various resistance movements, and from 1943 the US Office of Strategic Services took part in these operations. Besides sending agents to

[1] Major Edgar M. Howell, *op. cit.*, p. 199.

THE AIR FORCE IN GUERILLA WARFARE

occupied Europe, SOE and OSS supplied weapons, explosives and wireless sets to partisans in the three main theatres of war. The shortage of aircraft limited help to Western Europe at first, but during the peak period in 1944 France alone was supplied by 1,000 British and American aircraft per month, bringing in equipment for 50,000 men. In the Middle East supplies, as well as organizers and commanders, were first brought out in 1941 by forty aircraft from Cairo; later on Italy became the main supply base for help to Tito's partisans and in view of the large requirements a regular Air Force HQ, the Balkan Air Force in Italy, took over this task in 1944. In the third theatre, the Far East, SOE in Delhi sent supplies by air from 1941 on to Burma, Malaya, and Indo-China.[1]

The Red Air Force undertook similar missions. It had aircraft with the Balkan Air Force and dropped supplies to Communist partisans in Yugoslavia, Poland and Czechoslavakia; the latter also received officers with guerilla experience and about 400 partisans from Soviet Russia. Most of its aid was, of course, destined for its own partisans. Agents, partisan leaders, staff officers and NKVD functionaries were dropped and landed, sometimes intelligence reports were transmitted or wounded evacuated, but most of the flights were supply missions. According to German estimates Soviet partisans in the German Army Group Centre area received up to 150 supply drops a night in the summer of 1943[2], while Soviet sources state that in all 7,000 aircraft with supplies were sent to the partisans during the war.[3]

With very few exceptions the Allied air forces did not carry out reconnaissance flights for the partisans and they did not airlift partisan units. Only Tito's partisans had an air force of their own. Apart from the three German light bombers which they had managed to capture in 1942 and which the enemy soon discovered and destroyed on the ground, they had formed, with Soviet help, at the end of 1944 two air divisions which took part in the final partisan operations in northern Yugoslavia.

In the anti-partisan fight the efforts of the Luftwaffe were res-

---

[1] Cf. for particulars Brigadier-General Berry, 'Statement by U.K. Representatives', in *European Resistance Movements, op. cit.*, pp. 349 f.

[2] H. Teske, 'Partisanen gegen die Eisenbahn', *Wehrwissenschaftliche Rundschat*, October 1953, Heft 10, p. 473.

[3] Cf. Raymond L. Garthoff, *Soviet Military Doctrine*, Glencoe, Illinois 1953, p. 402, for reference.

tricted by the shortage of aircraft, but there is no doubt that the German High Command was fully alive to the importance of the air in anti-partisan warfare. Altogether seven functions were assigned to the Luftwaffe:

1. Assisting the ground forces in reconnoitring partisan movements, bases, camps, etc., through visual observation and aerial photography.
2. Destruction of enemy supply aircraft.
3. Operations against partisan bases, camps, etc.; support of the ground forces by bombing and aerial cannon and machine gun fire.
4. Supplying troops in case of logistic difficulties.
5. Dropping propaganda material.
6. Providing transport for higher commanders before and during operations.
7. Transport in special cases of parachute and glider-borne troops and dispatch of agents.

The Luftwaffe set up special air force commands in the principal partisan areas and they were in charge of the air force units detailed for anti-partisan warfare. Air support was as a rule given only in bigger engagements and that meant mainly in encirclement operations. We have already mentioned the Stuka formations assisting the ground forces in Operation *Frühlingsfest* in Soviet Russia, and in Yugoslavia the Luftwaffe gave support in many forms and sometimes in great strength. During the Third Offensive aircraft and tanks unsuccessfully tried to relieve the small German and Ustase garrison of Livno; in the Fouth Offensive German and Italian formations from five airfields gave close support, subjecting the partisans to incessant air attacks and trying to block the escape route by bombing a bridge; in the Fifth Offensive the Luftwaffe supplied forward troops, bombed Tito's positions and machine-gunned the partisans day after day; it gave similar air support during the Sixth Offensive; and in the Seventh Offensive Junkers 52 dropped parachute troops and brought in gliders for the attack on Tito's HQ, reconnaissance aircraft searched for him after his escape and Stukas harassed his force.[1] There is no reason to doubt that the Luftwaffe's intervention was as effective as air support of ground troops can ever be.

The Germans recognized that air support, in order to be useful,

[1] Cf. Brigadier Sir Fitzroy Maclean, *Disputed Barricade, op. cit.*, pp. 187, 203, 213, 222, 227, 249 and 259, and Vladimir Dedijer, *Tito Speaks, op. cit.*, p. 192.

must be massive, and therefore, instead of giving some support to all anti-partisan actions they gave full support to particularly important operations one at a time. It can also only be effective if the air force is constantly kept informed about the partisan situation and the progress of individual operations in which it is engaged. For this reason ground and air must constantly exchange information. The CO of the Luftwaffe in partisan areas, as well as the formations, therefore kept close contact with the anti-partisan ground forces and in anti-partisan operations liaison officers were appointed where this seemed useful. Before an operation flight commanders had to be supplied by the ground force commander with maps exactly indicating the situation on both sides, they had to be informed about any changes, and they had to be furnished with the operational plans of the ground commander. In order to ensure the constant flow of information during the operation, the forward elements of the ground troops were supposed, if at all possible, to carry wireless equipment. To avoid accidental strafing of their own troops, recognition and communication signals were laid down before each operation.

The Germans found in Soviet Russia and Yugoslavia, as did the British later in Malaya, that partisans regard the appearance of reconnaissance aircraft as the first sign of an impending action against them and frequently disperse at once. This occurred particularly in areas which were not often flown over, and air reconnaissance was therefore restricted in such cases. However, we have remarked before that developments in aerial photography now make it possible to survey the battle area without warning the partisans. The Germans did carry out air reconnaissance if the partisan detachments were large and therefore probably willing to fight it out, and also in search of dispersed partisan elements.

The Germans attached great importance to air reconnaissance. Photographs had, of course, to be interpreted by trained air force personnel. Sufficient copies had to be made of useful pictures which were distributed down to battalion level. Flying personnel was instructed with the aid of aerial photographs about the special characteristics by which partisan detachments and their installations could be recognized and, on the other hand, it was found desirable to instruct army officers in the interpretation of aerial photographs. Special attention was directed to the observation of partisan supply flights and the ground arrangements in dropping zones, since a lot could be learnt from them about partisan concentration areas, strength, bases and camps.

In view of the controversy about the effectiveness of bombing in anti-partisan operations it may be noted that the Germans found it really effective for the destruction of fortified bases and prepared camps and for attacks on partisan assembly points and battle concentrations; bombing of human targets could in the German view be suitably supplemented by low altitude cannon and machine gun fire if the defence was weak.

Attacks on village centres were not recommended, since partisan settlements are usually not located there but near the outskirts and gates of villages and in neighbouring forests. Attacks were made at dawn for choice because in daytime partisans are often away on their various missions. If possible, several villages were attacked at the same time, since partisan staffs frequently change their location and move to a neighbouring village.

The German Manual for *Warfare against Bands* contains one directive of particular interest: 'In special cases parachute or gliderborne troops may be required to achieve surprise in closing the encirclement ring or in blocking escape routes' (No. 138). We shall see later that this idea was subsequently used by the British and French.

We propose to turn now to the post-war theatres of guerilla operations in which the air force played its part, notably Greece, Malaya, Indo-China, Cyprus and Algeria. Since post-war guerillas never enjoyed air support, one must look to the anti-guerillas for any developments in air techniques. The most notable features are the introduction of helicopters into guerilla fighting and the closer integration of the air force into the war effort. These techniques were exemplified in Malaya in the summer of 1955 in Operation Unity, when Canberra jet bombers, Valettas, Pioneers, Austers, and last but not least helicopters, were deployed in the hunt for Chin Peg, the Secretary-General of the Malayan Communist Party, along the border of Malaya and Thailand. First, Valettas dropped parachute troops who prepared landing zones, then came Whirlwind helicopters with tractors for building airstrips, troops were then flown in by helicopter, which also transported troops and heavy stores into and out of the area during the hunt, and Sycamore helicopters stood by for the evacuation of the wounded. Austers took part in reconnaissance while the Canberras gave air support.

THE AIR FORCE IN GUERILLA WARFARE

Integration on this scale was not achieved without trial and error. Helicopters were first used in Korea, but mainly for observation, reconnaissance, liaison and the evacuation of wounded, and guerilla war seemed to deprive fixed wing aircraft of most of their traditional targets: there were few worthwhile objectives for strategic bombing and those there were were in Yugoslavia and Albania out of bounds for the Greeks, or in China and out of bounds for the French. Apart from that, guerillas are masters in camouflage and they were helped by the features of the countries in which they chose to operate. Even tactical air support appeared severely limited. Guerillas march by night and cannot be hit, and they frequently attack by night. Jungle and forests afford protection from observation and from the weapons of attack; 'the weight of rocket, cannon or machine gun attack in normal angles of dive is largely absorbed by the upper strata of foliage'[1] and thick jungle reduces their explosive effect; indeed, 'Field-Marshal' Kimathi has claimed that bombers killed only nine rebels in twelve months in Kenya.[2] Since the post-war guerillas had no air support, the anti-guerilla air force was never called upon to achieve supremacy in the air. Even as transports the air forces were often restricted; their bases, especially in Indo-China, were far away, airfields were few, supplies could not be dropped over jungle and forests, and to drop troops there appeared too hazardous—until the proper technique could be developed. If one thinks of the other air force handicaps, the guerilla terrain and the climate, it is almost surprising that in the end the air force could make such a remarkable contribution, especially to the tactics and techniques of anti-guerilla combat.

In the Greek civil war the air force had two tasks, viz., to fly independent missions and to give direct support to the ground troops.[3] The independent missions were of three types: reconnaissance in order to locate targets for the air force, armed reconnaissance against probable targets, and attacks on known targets. Possible targets, if they could be located, were troops on the march, in bivouac or concentrated for attack and defence, guerilla-held towns and their

[1] Wing Commander C. N. Foxley-Norris, 'The Use of Airpower in Security Operations', *Royal United Service Institution Journal* 1954, p. 555.
[2] Ione Leigh, *In the Shadow of the Mau Mau*, London 1954, p. 197.
[3] Cf. Colonel J. C. Murray, 'The Anti-Bandit War', *Marine Corps Gazette*, January to May 1954, and for the description above his very instructive article of May, pp. 52 f. All five articles deal with the civil war in Greece.

defensive positions. Direct support missions were reconnaissance and observation for ground troops and artillery, supply flights and attacks on targets in conjunction with ground troops. The government forces never fully realized the air potential. Their ground-air communication technique in particular was less well thought out than the American in the Philippines or the German in Yugoslavia; the number of wireless sets capable of calling for air support was entirely insufficient and they were not located with the assault troops but with brigade and division. The Greek Air Force, however, made such a considerable contribution in the final battles of 1949 when it was employed on reconnaissance and in attacks with bombs, rockets and napalm in support of the ground forces' advance that Colonel Murray, who took part in that war, concludes that the victory in Vitsi 'was the result of the effectiveness of artillery and air support'—and the guerillas' failure to defend a key position.

Reconnaissance and aerial fire support were thus the two main functions of the Greek Air Force. Naturally, an air force cannot win an anti-guerilla war on its own. Also, as Wing-Commander Foxley Norris has pointed out, it is no longer possible for an air force to maintain peace without the use of large ground forces as the RAF did between the wars on the North-West Frontier, in Iraq, Aden and Somaliland.[1] In guerilla wars it can only play a supporting role. But in doing so it has made possible ground operations which could not otherwise have been carried out, it has made it possible greatly to improve the efficiency of practically every type of anti-guerilla ground action, it has increased the combat soldier's 'output per man hour' and it has made his life easier. The following survey will try to substantiate these claims. We shall not describe air force actions by campaigns but by indicating kinds of ground operations which the air force usually supported.

Transport is required for all ground operations. In Indo-China the greater part of the air effort consisted in transport support. Aircraft were the sole means of supply for all independent defence posts in Laos; Dien Bien Phu and Na Sam were always supplied by air and so were a great many other posts and forts, even those within normal reach of road convoys, because they were endangered by ambushes and required considerable guard personnel. The garrisons of these posts were also relieved and reinforced by air. In Malaya too trans-

[1] *Loc. cit.*, p. 556.

port support took first place among air missions. Pioneer light aircraft and helicopters kept up communications with the jungle forts and delivered the supplies; transport for visiting doctors and police officers and the exchange of garrisons were also provided by the RAF. If troops on the march had to be supplied, as frequently happened, they selected a dropping zone and marked it by fire, an identification letter, or fluorescent panels, and pilots found the DZ by map reading and time runs on a compass bearing from a known datum point.[1] With the advent of helicopters they took on the task of evacuating casualties in most theatres of guerilla war. In Malaya troops called the base for this purpose by R/T, an Auster aircraft would then contact the patrol and guide it to the nearest clearing; the ground was then prepared and the helicopter flew in.[2]

Many air transport missions have been flown in direct support of ground operations, and in this latter field the air forces have been called upon to fulfil the following transport and other tasks:

1. Reconnaissance before action. When the jungle war started in Malaya it was found that the existing maps of the country were inadequate and the many gaps had to be filled by aerial photography; the task was completed early in 1953. Even after this date aerial reconnaissance ranked second in importance to transport in the work of the RAF. That it did not figure equally high on the list of air force priorities in Indo-China was all the more surprising because ground intelligence was inadequate owing to the attitude of the people; the situation might therefore have called for a stronger air effort. On the British side visual reconnaissance was considered the most reliable type of intelligence, and photographs as well as other information were checked by visual reconnaissance. This mission was assigned in Malaya to army pilots flying Auster aircraft. Each pilot was allotted a particular area with which he became thoroughly familiar and where he could spot any change which might indicate the presence of terrorists.[3] Helicopters were used on reconnaissance there for certain detailed tasks such as pinpointing a camp whose location was only approximately known or spotting a river crossing.[4] This last task

---

[1] Cf. for the above Flight Lieutenant D. R. Seeth, 'The Employment of Air Power in Malaya', *Indian Air Force Quarterly*, October 1954, vol. iv, No. iv, pp. 38 f.

[2] Cf. Flight Lieutenant D. R. Seeth, *loc. cit.*

[3] Cf. Group Captain K. R. C. Slater, 'Air Operations in Malaya', *Royal United Service Institution Journal* 1957, p. 380.

[4] Group Captain K. R. C. Slater, *loc. cit.*, p. 380.

PARTISAN WARFARE

meant in Malaya finding a few boats, while pilots in Indo-China often had to search for an under-water path formed by planks—a method which Soviet partisans had used before to make their camps in deep swamps appear inaccessible.

Aerial reconnaissance also tried to spot bands on the move. In Algeria and Cyprus helicopters enabled the CO to carry out his own reconnaissance before an engagement and also to visit his sub-units. In Malaya it became customary for the patrol commander shortly before the beginning of an operation to be flown over the intended scene of action and the pilot briefed him during the flight on ground conditions, selection of dropping zones, patrol routes etc.[1]

A special kind of observation was carried out in Indo-China by French liaison planes, which would escort road convoys, try to spot ambushes and call for reinforcements in case of need. This technique was not considered successful; by the time help had arrived the Vietminh had usually disappeared.[2] The moral is obviously that escort aircraft must be equipped with cannon and machine guns so that they can attack before the enemy vanishes.[3] A rather different duty was assigned to slow aircraft in Cyprus. They too flew above road convoys in order to minimize the danger from mines and ambushes and tracked a man until troops were brought up to the spot where he had gone into hiding.[4]

We have so far discussed air reconnaissance in preparation for army action and we shall later deal with air force reconnaissance for its own purposes.

2. *Support of ground troops in action.* Air support has been given to army units of greatly differing strength, ranging from the small patrols of a few men to large garrisons of many thousands. The following examples may be considered typical; we start with air support of patrols.

When the guerillas withdrew farther and farther into the jungle in Malaya it became evident that patrols could hardly penetrate so deeply unless supplies were dropped to them, and these drops became a regular mission of the Valetta force. The mountain patrols in Kenya and observation posts in Cyprus were similarly provided for.

[1] Cf. *Eagle*, New Year 1954, p. 5.
[2] Colonel J. F. McQuillen, 'Indochina', *Marine Corps Gazette*, January 1955, p. 51.
[3] Cf. also Wing Commander C. N. Foxley-Norris, *loc. cit.*, p. 557.
[4] Cf. Dudley Barker, *loc. cit.*, p. 187.

Also, instead of marching for days in the jungle, patrols in Malaya could be air-lifted by helicopter to a place nearer to their objective, and similarly, observation posts in Cyprus could be sited higher than ever before. In Algeria sweeps could be carried out in inaccessible terrain: preceded by an observation aircraft which marked the landing zone, four helicopters would take up thirty-two men in all and after landing them go back several times for more.[1] In Malaya, in 1951, when the British High Commissioner was murdered, the RAF and artillery put down a curtain of fire in order to make the bandits move into a position where they could be ambushed by the waiting troops.[2] In Cyprus helicopters carried out the initial lift of troops in search operations[3] and achieved complete surprise, whereas troops marching to their objective can hardly ever hope to take the enemy unawares. In Operation Mare's Nest, in the Troodos Mountains of Cyprus, the search for hideouts started with eight helicopters landing soldiers on top of ridges in dominating positions and before the terrorists had a chance to move out of the area a network of sixteen posts, inaccessible by road, had been established. The area was kept under observation by air and ground patrols in daytime, and by night ambush parties covered every escape route. 1,700 men in all took part in the operation. By the use of helicopters the observation posts could be established in the most advantageous places much deeper in the mountains than ever before, and supplies were assured. The operation was called off when truce negotiations appeared possible.[4]

Operation Termite was one of the successful encirclement operations; it lasted from July to November 1954 and was mounted with strong air support in the deep jungle of Perak. It appeared from intelligence reports that a considerable number of terrorists, supported by aborigine informers and patrols, were concentrated in the area; the terrorists believed that no security forces could reach them undetected. Surprise could only be achieved with the help of the RAF. The operation started with accurate bombing of selected targets by the RAF and immediately afterwards three squadrons of the 22nd Special Air Service Regiment made a parachute jump and blocked

[1] Major David Riley, 'French Helicopter Operations in Algeria,' *Marine Corps Gazette*, February 1958, p. 24.
[2] Cf. Harry Miller, *Menace in Malaya*, London 1954, p. 195.
[3] Cf. Major P. D. F. Thursby, 'Helicopter Operations in Cyprus', *The Suffolk Regimental Gazette*, Summer 1957, p. 17.
[4] Compare for the above operation the article in *Soldier*, vol. 15, March 1959, pp. 14 f.

the escape routes. Ground forces then closed in. During the lengthy operation the majority of the troops were supplied by air and troops were frequently lifted by helicopters, which also evacuated casualties.[1] This show of force considerably impressed the aborigines and paved the way for winning them over. At the same time the problem always present in encirclement operations, viz. how to get the troops into position close to the target area without giving the guerillas time to evacuate, had been solved by the use of air-power. The French used the same technique with success in Algeria.

Two of the squadrons of the 22nd Special Air Service Regiment which had taken part in Operation Termite later went to Oman and, supported by Life Guards and local forces, in a brilliant action took the Jebel Akhdar mountain range and defeated the Omani Liberation Army. But here the RAF's share in the honours was limited, its activities being restricted to visual and photographic reconnaissance, patrólling and dropping supplies.[2]

But again, the air force played a leading part in the successful defence of Na Sam, in the Thai country, in 1952. The fortress had always been air-supplied and it was, as the reader will recall, the last remaining post in the French defence line. It was held by a garrison of 20,000 men. The Viet-minh pressed hard. The air force brought reinforcements just in time. The Viet-minh encircled a strong point which dominated the area and possession of which would have allowed them to control the fortress, but the French Air Force broke the encirclement ring by a heavy bombardment and drove the enemy back, and parachutists then succeeded in retaking the position.[3] Unlike Dien Bien Phu, the Na Sam airfield had been incorporated into the defence system of the fortress. The battle for Dien Bien Phu was lost, as has often been pointed out, when the airfields were lost; at Na Sam decisive air support was possible because the airfield was retained.

In addition to all these missions in support of ground troops in action, air forces have blocked partisan escape routes by aerial fire or bombing of bridges; they have succeeded, on one occasion at least,

---

[1] Cf. *Federation of Malaya, Annual Report, 1954*, p. 415.
[2] For a report of the campaign see 'Rebels Surprised—Brilliant but Little Known British Desert Action', London *Times*, April 9, 1959.
[3] Cf. Lieutenant R. Cauchetier, 'Na-Sam', *Forces Aériennes Francaises*, Indo-China 1953, pp. 569 f.

THE AIR FORCE IN GUERILLA WARFARE

in bringing up reinforcements to ambushed troops;[1] in Cyprus helicopters transported troops quickly on receipt of hot tips;[2] in Algeria they frequently flew the commander of the ground troops over the area during the operation; they have delivered rocket and dive bombing attacks to keep the guerillas moving so that the ground troops had a better chance of finding them, and they have assisted the artillery by acting as spotters and markers.

The air forces have also executed a great number of independent missions. They have carried out reconnaissance in preparation for their own attacks and subsequently bombed camps and sprayed crops in the deep Malayan jungle or in Indo-China. But it became increasingly more difficult to locate camps and if they were found they were usually deserted,[3] cultivations became smaller and better camouflaged and if they were not in partisan areas it was difficult to decide whether they belonged to friend or enemy. The technique of marking targets and precision bombing by day and night was developed,[4] but in view of the scarcity of targets bombing took last place in the order of importance in Indo-China and Malaya. Bombing of area targets was, however, considered important in Kenya. As General Erskine has said, 'in circumstances of this kind you have to be satisfied with indirect results such as an increase in surrenders, a drop in Mau Mau morale, prisoners' reports, and similar evidence. It is seldom that you can report actual casualties immediately after an attack.'[5]

In the psychological warfare campaign, particularly in Malaya, it fell to the air force to deliver the goods. Leaflets were dropped from aircraft and the now famous voice aircraft, playing back messages from surrendered terrorists, made their appearance. The range of these messages was one mile at 1,500 feet and under ideal conditions an aircraft could cover 100 square miles in an hour.[6]

In the various air activities helicopters played an increasing role. In the guerilla theatres they have been used for command and liaison flights, reconnaissance, troop transports, evacuation of casualties, supply transports, artillery spotting, directing air strikes and spraying

[1] Report in London *Times*, January 28, 1957.
[2] Cf. Major P. D. F. Thursby, *loc. cit.*, p. 17.
[3] Group Captain K. R. C. Slater, *loc. cit.*, p. 380.
[4] *Ibid*, p. 379; also *The Eagle*, Summer 1955, p. 18.
[5] General Sir George Erskine, 'Kenya—Mau Mau', (Lecture) reprinted in *Royal United Service Institution Journal* 1956, p. 17.
[6] Flight Lieutenant D. R. Seeth, *loc. cit.*

crops, and in Algeria they were used for the first time as troop assault vehicles.[1] They have taken along tractors for building landing strips and they could probably carry tanks and field guns. In Cyprus helicopters were not at first available, but when they were, they made it possible to carry the war practically right into the enemy's camp and the consequential new techniques developed by General Darling would have resulted in a clear British victory had not the truce intervened.

As far as the air effort as a whole is concerned, its greater part in Malaya and elsewhere, as we have stated before, was in transport support. It may not be a glamorous task but its contribution to the efficacy of anti-guerilla fighting cannot be over-estimated. In the words of Group-Captain Slater, 'supply dropping (in Malaya) by the Valetta force, coupled with the troop-lifting and casualty evacuation by helicopter, have combined to multiply the number of troops and police deployed on productive jungle patrols by a factor of *no less than four*'.[2]

It is obvious that the same high factor cannot be achieved in every theatre of guerilla war; in the jungle progress on the ground is particularly slow. But everywhere air transport is quicker than marching, everywhere air-supplied troops can stay out for weeks and need not return after days, the troops need not carry their wounded for long and patrols need no longer return as they might have had to, after detailing men as escorts and stretcher bearers. It is thus evident that in the effective use of air-power the anti-guerillas have a means to lessen or overcome the handicap under which they all suffer: the lack of sufficient combat troops. They can now make more effective use of what they have.

Finally, the element of surprise has always been on the guerillas' side. Air-power has now restored the balance. The anti-guerillas, through air-power, have a greater chance of forcing the enemy on to the defensive. 'The toughest thing for us,' said the most experienced of all Soviet partisan leaders, Major-General Kovpak, 'had been the defensive battles forced upon us by the enemy. Only in these battles had he been able to take advantage of his numerical and technical superiority,'[3]—a most important point.

[1] Cf. Major David Riley, *loc. cit.*, p. 22.
[2] Group Captain K. R. C. Slater in the already quoted article in the *RUSIJ*, p. 382. The above italics are ours.
[3] *Our Partisan Course*, op. cit., p. 81.

THE AIR FORCE IN GUERILLA WARFARE

To sum up, if the air force can no longer go it alone, it can at least make a great contribution towards victory.

It appears that the pattern for the use of air-power in anti-guerilla warfare has been set, and that no great changes can be expected in the future. There might be developments in helicopters—such as improved air-ground communications for the airborne commander or rockets and machine guns for self-defence[1]—and anti-guerillas might receive a more generous allotment of aircraft. Whatever the function of aircraft under atomic warfare conditions, the types suitable for anti-guerilla warfare must never be allowed to go out of production.

It must not be overlooked, however, that so far the air force has had it all its own way; only over the Russian front did anti-guerilla aircraft encounter opposition in the air and only in Indo-China were they exposed to guerilla anti-aircraft guns. In a future war guerillas might have an air force, anti-aircraft and radar at their disposal. But if the guerillas were fighting on their own, a war under such conditions would no longer be a guerilla war but a conventional one, and if they received this support from their own or a friendly army, the anti-guerilla air force would have a few more problems on its hands; these again are mostly problems of a conventional war and outside the scope of this book.

How was ground-air co-operation organized at top level? The Malayan arrangement is not of general interest because it was designed to meet a special situation inasmuch as, for political reasons, control of the ground forces was decentralized, while the RAF was centralized under Air HQ. Yet a suitable compromise solution was found there.[2] In Kenya control of both Services was centralized and so was planning. In a Joint Operations room Army and RAF exchanged information, decided priorities and directed operations.[3] In Algeria, where the air force is partly under Army control and partly under Air Force control, there are three corps areas, and each corps area is subdivided into four divisional zones. Control is decentralized. Bombers, long range reconnaissance aircraft, including some helicopters, and heavy fighters are under the control of the Tactical Air Group with each corps. At divisional level the tactical air officer with division has

---
[1] Cf. Leading article in London *Times* of April 4, 1961.
[2] Cf. Group Captain K. R. C. Slater, *loc. cit.*, p. 386.
[3] Cf. General Sir George Erskine, *loc. cit.*, p. 17.

PARTISAN WARFARE

under his command one squadron of SNJ aircraft based in the zone for armed reconnaissance missions, and each divisional commander has an army observation aircraft platoon under his own command.[1] Finally at the bottom of the scale, light helicopter squadrons of both services are assigned in units of one or two to the widely dispersed garrisons.[2]

The British view has always been that the air is indivisible and command must be centralized. Since a specific mission might involve a number of different types of aircraft, the advantages of decentralization do not seem obvious.

In anti-partisan operations airfield security is of great importance. The 1944 Field Service Regulations of the Red Army laid down that in attacks on an enemy aerodrome or landing strip 'one determines the precise location of the aircraft, the approaches to the machines and to the fuel and ammunition depots. In working out a plan of attack, special attention should be given to wiping out the guards located in barracks and tents. The covering groups carrying out the destruction of the aircraft and fuel should be reinforced. The cover should draw to itself all the fire of the guards in the guard house or in the barracks. One may also set fire to and destroy hangars, machines, bomb depots and fuel dumps by the fire of anti-tank rifles, rifles, armour-piercing incendiary bullets, grenades, mines, incendiary mixtures, and thermite compositions.' For the attack the detachments were usually divided into three sections, the assault group, the demolition group, and the reserve group.

The Viet-minh prepared their attacks on French air bases with much greater thoroughness and subtlety. Airfields and transport planes were a priority target, since the French war effort depended on air transport; deprived of reinforcements and supplies by air the French defence system would have collapsed. Viet-minh commandos were specially trained for months for this task, and they were supplied with all the information desired by Viet-minh agents; the agents received their information from the fifth column and the reader will recall the Viet-minh statement in their Manual that 'it is easy for us to find "intermediaries" and to create cells among them.... They skilfully co-operate with us'. It is therefore not surprising that the

---

[1] Cf. Colonel Victor J. Croizat, 'The Algerian War'. first published in the *Marine Corps Gazette* and reprinted in *An Cosantóir*, 1958, p. 18.

[2] Cf. Major David Riley, *op. cit.*, p. 23.

THE AIR FORCE IN GUERILLA WARFARE

commandos managed to slip in undetected, even by the tracker dogs. It also appears that the aircraft were insufficiently guarded and that the guards were not trained to react promptly in the face of a surprise attack.[1] The attack itself proceeded on the usual lines.

The most important partisan sabotage objectives are, of course, aircraft on the ground, and only if the saboteurs have sufficient time will they destroy radar and wireless installations and fuel dumps.[2]

The defence of air bases is always a difficult matter, as the SAS and LRDG exploits show. A forbidden zone round the airfield is of some help; indigenous personnel may have to be dispensed with. The French air bases in Indo-China were guarded by three infantry companies to each base and they provided the entrance guards, the guards for the aircraft, the base patrols—on foot or in half-track vehicles—and the watch tower guards. The towers were located round the perimeter at about 450 yards interval or less, if necessary, and equipped with searchlights. The perimeter itself was secured by a continuous wire obstacle.[3] But whatever the arrangements are and however perfect they may appear, the successful defence of a base depends in the last resort on the training and the determination of its personnel.

[1] Cf. Général Ch. Lauzin, 'Opérations en Indochine', *Forces Aériennes Françaises*, March 1955, p. 464.
[2] Sabotage techniques are described by Hauptmann H. von Dach Bern, *op. cit.*, p. 76, and for the destruction of jet aircraft see illustration p. 77.
[3] Cf. Commandant Mitaux-Marouard, 'La défense des bases aériennes en Indochine', *Forces Aériennes Françaises*, Indochine 1953, pp. 596 f.

CHAPTER 9

GUERILLAS AND NUCLEAR WARFARE

It is obvious that in this so-called atomic age guerillas are as active as ever and that the mere existence of nuclear weapons has not made guerilla warfare obsolete. That guerillas will have their uses before any nuclear weapons are exploded in a nuclear war does not require discussion either. But will they still have any part to play after nuclear weapons have been used? It appears unlikely that such weapons will be used against independent guerillas and we will discuss here only the position of guerillas fighting with their own troops, viz. of auxiliary guerillas.

The supreme maxim for the conduct of a nuclear war is to force the enemy to concentrate his troops while never forming concentrations oneself. Guerillas need not change any of their tactics to comply: they never concentrate, except for attack, and then only for the shortest possible time. In fact, it looks as if nuclear warfare offers a number of advantages to guerilla forces, viz.:

1. If the enemy troops are more widely dispersed, their installations must also be more widely dispersed, and this means more targets for the guerillas.[1]

2. Dispersed enemy troops will be less capable of mounting large-scale anti-guerilla operations. The guerillas have therefore a greater chance of survival than hitherto.

3. In this kind of warfare there may be more and wider areas left unoccupied by the enemy, and if this is so, the guerillas will have a greater chance of development.[2]

4. The guerillas would not have to fear a nuclear attack on them-

---

[1] Cf. Gene Z. Hanrahan, 'Guerilla Warfare', *Marine Corps Gazette* 1956, p. 31, and Roy Farran, *Operation Tombola*, London 1960, p. 204.
[2] Valdis Redelis, 'Partisanenkrieg', *Die Wehrmacht im Kampfe*, Heidelberg 1958, p. 52. Cf. also Colonel Demange, 'La Guérilla', *Revue Militaire Générale*, May 1960, p. 565.

GUERILLAS AND NUCLEAR WARFARE

selves. In the first place, they are too small a target for a nuclear attack, and secondly, they operate behind the enemy and he would hardly endanger his own hinterland by exploding nuclear weapons there.

Now all this presupposes that the attacker in a nuclear war will move out of his own country and occupy part of his opponent's territory. But he might equally well decide to stay in his own country, secure his frontiers, and fire his nuclear weapons from his present land bases, the sea and the air into his opponent's country. In this case the opponent's guerillas would have nobody to fight against. But if the attacker moves out, Mao's old dictum applies that wherever the enemy can go, guerillas can go too. Whatever the devastation wrought by nuclear warfare, if the enemy survives the guerillas can survive too and carry on guerilla war.

The guerilla targets will be the same as before, with, in addition, the paraphernalia of nuclear war. The guerillas will have to reconnoitre the location of atomic weapons and, if possible, destroy or sabotage them. They must report to their troops what use the enemy intends to make of atomic weapons. They must report enemy troop concentrations, since these might be a suitable target for the nuclear weapons of their own side.

In a nuclear war the partisans will have two essential tasks. In the first place they must prevent the use of nuclear weapons by the enemy; or if he has used them, the partisans must prevent him from exploiting the situation. In the second place the partisans must assist in providing favourable opportunities for the use of nuclear weapons by their own side and, when these weapons have been used, the partisans must help to exploit the situation.

The partisans will prevent the enemy from using his nuclear weapons by destroying them and attacking the personnel who operate them. But it must be realized that the partisans' chances are limited. Long-range nuclear missiles are not necessarily fired from land bases and even if they are, the bases are far behind partisan territory, and so are the airfields of the air force detailed for nuclear strikes. The partisans can at best only interfere with tactical nuclear weapons. Again, one must not overrate the partisans' capabilities when it comes to preventing the enemy from exploiting the situation. In all probability he will use tanks for this purpose and if they assemble close behind the front line the partisans will have little opportunity for sabotage. But if the enemy concentrates his tanks before the nuclear attack the partisans must report accordingly and enable their own army to destroy them, possibly with tactical nuclear weapons; and if the enemy keeps his tanks dispersed in his rear the partisans can disable them themselves.

Once the enemy has fired his nuclear weapon and opened a breach in the front, the partisans must help to prevent an enemy breakthrough. Without venturing too far into the field of speculation, we may conjecture that in such a situation an attempt will be made to close the gap with tanks, and if this happens there will be a race to the gap by the tank forces of both sides. In these circumstances the partisans will do a splendid job if they succeed in delaying the enemy tanks, by destroying bridges, laying ambushes and controlling air and artillery attacks.

The partisans will help to create favourable opportunities for the use of nuclear weapons by their own side if by cutting other communications, they compel the enemy to move his troops on one or two roads and thus to concentrate his troops in a limited area. Partisan intelligence should report not only actual enemy troop concentrations but also opportunities for the partisans' own troops to induce or compel the enemy to form concentrations.

Finally, partisans will help their own troops to exploit a nuclear attack by preventing the enemy from closing the resulting gap in his front line. They will block roads leading to the gap and attack rail and motorized transport in the area. They will also report the results of the nuclear attack and the enemy's counter-measures.

After an enemy nuclear attack the partisans may well be the only opposing forces in the area. This does not make the position more dangerous for them, because the enemy will exploit the situation and move his troops forward. Auxiliary partisans will probably not be sufficiently trained or equipped to draw the enemy's attention upon themselves in such a situation in order to give their troops time to seal the breach.

It therefore appears that a nuclear war would offer the partisans a number of advantages, but if nuclear weapons were about to be fired or had been fired by either side, the partisans could give only limited assistance. Their main contribution in this case might well be in the field of intelligence.

CHAPTER 10

THE TREATMENT OF GUERILLAS AND THE POPULATION

A. *The Treatment of Guerillas*
We propose first to look at the legal status of guerillas. If they fight against their own country they are subject to the penalties of their *domestic* law and every country has provisions for the punishment of those guilty of treason, taking part in armed revolt, or whatever the case may be.

Guerillas who fight against a foreign invader are subject to *international* law. International law does not consider participation in a guerilla war as an offence. However, The Hague Convention on Land Warfare enumerates certain rules. Guerillas who comply with them are lawful belligerents and, if captured, must be treated as prisoners of war. Those who do not comply with the rules are not lawful belligerents; they are called *francs-tireurs* and the enemy may deal with them as he thinks fit. The rules set out in Article 1 of The Hague Convention read:

'The laws, rights, and duties of war apply not only to armies, but also to militia and volunteer corps fulfilling the following conditions:

1. To be commanded by a person responsible for his subordinates.
2. To have a fixed distinctive emblem recognizable at a distance.
3. To carry arms openly.
4. To conduct their operations in accordance with the laws and customs of war.'

Practically all guerillas of the last war were *francs-tireurs*, but it was recognized by one of the American Military Tribunals in Nuremberg that certain partisan units in Yugoslavia and Greece had fulfilled the conditions of Article 1.[1] In the British view it must be

[1] Judgment in the 'Hostage Case', Case No. VII in the *Subsequent Trials at Nuremberg*.

proved that somebody has been acting as a *franc-tireur* and 'this presupposes some form of trial'.[1]

However, not all who fulfil the four conditions of Article 1 are lawful belligerents. They must also be in the service of a *de jure* or *de facto* State or an institution with some similar characteristics.[2] If the State has capitulated or if its armed forces have been completely defeated, the territory completely occupied and the government dissolved, the partisans of that country are not lawful belligerents—unless they have formed a *de facto* government exercising control over portions of the old State and its inhabitants.

What it amounts to is this: in the last war, after the defeat by Germany of a number of European countries, only the Soviet partisans whose State still existed, and Communist partisans who had set up a government, could thenceforth aspire to be considered lawful belligerents, provided, of course, they also fulfilled the four conditions of Article 1. The result is rather curious.

Tito's partisans were not lawful belligerents prior to November 26, 1942, when he set up the National Liberation Committee of Yugoslavia.

It is doubtful whether they then qualified as lawful belligerents, since Tito then held no territory, or whether such a claim was justified only after Tito's capture of Belgrade in October 1944. But from this time at the latest they were lawful belligerents.[3]

Yet the guerillas of Mihailovitch, who was Minister of War in the Royal Yugoslav Government, were never lawful belligerents—the King and his government had left the country, the Army had capitulated, and the country was occupied.

The state of the law may not seem satisfactory, but it must not be overlooked that when the rules were devised revolutionary wars were of little significance. In most cases, at any rate, guerillas are not entitled to the protection of international law. How, then, should they be treated?

It seems pertinent to inquire first what treatment guerillas mete out to their opponents in the field and to the population.

---

[1] Transcript in the 'Court-Martial case against Field-Marshal von Manstein', p. 3144.

[2] Cf. August von Knieriem, *Nürnberg, Rechtliche und menschliche Probleme*, Stuttgart 1953, pp. 380 f; q.v. also for references. Finally cf. Article 4 of the Geneva Convention 1949.

[3] A. von Knieriem, *op. cit.*, p. 382, states that it may be possible to regard Tito's forces, 'from a rather late moment on', as lawful belligerents.

TREATMENT OF GUERILLAS AND POPULATION

The well-known guerilla methods of dealing with the population are persuasion and terror, both reinforced by guerilla warfare. Not all guerilla movements use these methods, of course, and most purely patriotic movements never use the terror weapon.[1] But revolutionary guerillas find these methods indispensable, and propagandists, agitators, police, murder squads and guerillas are the agencies which put them into effect. The process is known as persuasive reasoning.

The revolutionary guerillas' main task lies in the field of persuasion. They too, it is true, have to deal 'blows to the traitors and collaborators who undermine the army and the people'[2] but they are kept out of terror activities as much as possible because 'the army must be at one with the people and be regarded by the people as their own',[3] thus securing an uninterrupted flow of army volunteers. According to Mao, in establishing the principle of unity between the army and the people, three tasks fall upon the former: 'enforcing such discipline in dealing with the masses as prohibits the army from violating even in the slightest degree the property rights of the people, carrying out propaganda among the masses and organizing and arming them, lightening the financial burden on the people.'[4]

The details of Mao's teachings are filled in by the Viet-minh Manual:[5]

'What one must do when stationed in billets:
'Propaganda among the population (involves):
'—Forbidding the soldiers billeted in a private house to walk about disorderly, to sing or to make a loud noise, to take anything away without the owner's permission;
'—Informing oneself about the regional customs and habits, the life of the people and their feelings towards the resistance; dispelling the uneasiness of the population, uniting in regional cadres in order to make propaganda for the (Viet-minh) government. Stay for choice with the poor peasants and people, try to understand their problems and help them in their daily work;

[1] For the question when terrorism by insurgents is successful see Brian Crozier, *The Rebels*, London 1960, pp. 159 seq.
[2] Mao Tse-tung, 'Interview with the British Correspondent James Bertram', in *Selected Works, op. cit.*, vol. ii, p. 96.
[3] Mao Tse-tung, *On the Protracted War*, loc. cit., p. 240.
[4] Mao Tse-tung, 'Interview', etc., loc. cit., p. 96.
[5] *Bulletin Militaire*, 1955, p. 451.

'—Taking part in the meetings and activities of the groups and the popular guerillas. If possible, organizing study groups together with the popular guerillas (in this way troops and guerillas can inform each other about their personal experiences and help each other in their studies);

'—In case of differences with the population, immediately organizing a recapitulatory control meeting in order to maintain the spirit of unity.

'To sum up, wherever troops are billeted they must follow President Ho Chi-minh's advice:

"Gain the affection, the trust and the esteem of the population".'[1]

Guerilla warfare strengthens the impact of propaganda. The guerillas in the field not only demonstrate that final victory is possible but they also win the people by their successes, their sacrifices and sufferings.

How, then, do guerillas treat their opponents?

Most guerilla forces do not take prisoners; even if they want to, they would not have facilities for keeping them. Castro took a different course: he disarmed captured government troops, asked them to fight for his cause and released unharmed those who did not want to.[2] Some Viet-minh units treated their prisoners honourably, and others did not. Conditions in the PW camps were horrible, brain-washing and communist indoctrination were applied in accordance with Mao's precepts. Mao expressed his policy in this way:

'We shall give lenient treatment to captive Japanese soldiers and certain captive Japanese officers of lower ranks who fought under coercion; we shall neither insult them nor browbeat them, but explain to them the harmony between the interests of the people of the two countries, set them free and let them go home. Those who do not want to go home may work in the (Chinese Communist) Eighth Route Army.'[3]

---

[1] For the Viet-minh organization of each village, district and province under their control cf. Capitaine André Souyris, 'Un procédé efficace de Contre-guérilla', *Revue de Défense Nationale* 1956, pp. 686 f. The inhabitants are organized by occupations as well as by age and sex, thus assuring a twofold supervision.

[2] Dickey Chapelle, 'How Castro Won', *Marine Corps Gazette*, February 1960, p. 39.

[3] In 'Interview' etc., *op. cit.*, p. 98.

TREATMENT OF GUERILLAS AND POPULATION

Mao has explained his motives elsewhere:

'The method for reducing it (the superiority of the Japanese army) is mainly to win over the Japanese soldiers politically. Instead of hurting their pride, we should understand and humour it and, by giving good treatment to prisoners of war, induce the Japanese soldiers to awaken to the Japanese rulers' anti-popular policy of aggression.'[1]

Mao applied the same policy to Chinese Nationalist prisoners of war, not without success.[2]

Can the anti-guerillas draw any conclusions from the foregoing as to how they should treat captured guerillas?

If the guerillas have won 'the affection, the trust and the esteem of the population' it will become all the more eager to support them if they are harshly treated by the enemy. This applies particularly if the guerillas themselves treat their enemy leniently. As far as the guerillas are concerned, they know the risks when they join and they can only become more determined to fight to the last if they cannot expect leniency from the enemy. And disgust at the treatment meted out by the enemy, if properly exploited by guerilla propaganda, will win them support even from those of their countrymen who have not hitherto regarded them with affection.

It is not widely known that the Germans in the last war finally came to the same conclusions and tried to make a drastic change in their policy. At first not only partisans were treated as such and hanged—as international law permits—but also partisan suspects, so-called partisan supporters, and partisans 'by definition'.[3] But when in the face of these and other measures partisan support by the population became ever stronger, the Germans promised to treat them as prisoners of war if they would 'come out of the hiding places'. The 1944 *Manual*, however, went much further than that and far beyond the requirements of international law. In its paragraph 163 it laid down the following:

'All bandits in enemy uniform or in civilian clothes who are cap-

---
[1] *On the Protracted War*, loc. cit., p. 229.
[2] For the shock treatment and brain washing technique applied see Colonel Francis F. Fuller, 'Mao Tse-tung, Military Thinker', *Military Affairs* 1958, p. 145.
[3] Dixon/Heilbrunn, *Communist Guerilla Warfare*, op. cit., p. 87.

tured in combat or surrender in combat are in principle to be treated as prisoners of war. The same applies to all persons who are found in the immediate battle zone and have to be regarded as band helpers, even if their participation in combat cannot be proved. Bandits in German uniform or in the uniform of an allied army will be shot after careful interrogation if they have been taken prisoner in combat. Deserters, regardless of the clothes they wear, must in principle be treated well. The bands should be made aware of this.'

The Germans had learned by bitter experience that harshness stiffened partisan resistance and that if anything could weaken it, it was the more lenient treatment of partisan prisoners. The Germans had learned the lesson too late but that cannot detract from the validity of the principle.

There is one further argument against the use of harsh methods against partisans: if they know that partisan prisoners will be shot or hanged, they will not believe in the sincerity of an amnesty promise. Yet an amnesty, granted at the right time, can considerably weaken the partisan movement and shorten the war, and we have already mentioned the success which amnesties achieved in Malaya and Kenya. One should not, without very good reasons, deprive oneself of this advantage. The fact that under international law partisans can be killed after some sort of trial is no good reason for killing them.

All this, it should be stressed, is meant to apply to the rank and file only; ringleaders and assassins must be treated differently.

It may be felt that things will be made too easy for the guerillas if they are treated as prisoners of war, but there seems to be no justification for thinking that more might join up as a result. Few soldiers can have been induced to volunteer by the knowledge that if taken prisoner they will be prisoners of war, and the same probably applies to guerillas too. On the contrary, the prospect of decent treatment seems to increase the guerilla surrender rate: in Malaya terrorists tried to find out from the population whether they would on surrender really be as well treated as the Psychological Warfare Section of the Director of Operations' Staff made out in its propaganda.[1]

In many cases decent treatment by itself will make law-abiding citizens out of unlawful belligerents; in some cases it will help to turn them into pseudo-gangs. There is no need to add anything to what

[1] See Federation of Malaya, *Annual Report* 1956, p. 441.

we have said in Chapter 4 to underline the importance of this point. Surrendered enemy personnel in overseas territories are usually kept in detention and rehabilitation centres, whence they can be released even while the emergency is still on.

In recent guerilla wars propaganda has played a considerable part. Anti-guerilla propaganda is most effective when it can take as its line: 'Fight on and you will be killed; surrender and you will be safe', spiced with such items as defection of guerilla leaders, the worsening of the guerillas' supply situation, their food shortages, and so on. It is important not only that the information should be correct, but also that the guerillas should be able to check at least some of it, and as many names as possible—of surrendered, killed and captured terrorists, the location of engagements, etc.—should be given.

The aims and means of psychological warfare in guerilla war are the same as in any other war. We have already mentioned the voice aircraft as a particularly successful medium for dissemination of propaganda, and broadcasts of talks with surrendered bandits and mouth-to-mouth propaganda by specially employed former guerillas have also proved effective.

B. *The Treatment of the Population*
The population in the country's own territory is subject to the domestic laws, including the emergency regulations. The latter usually give the government power in revolutionary wars to order detention, declare protected places, control movements on roads, disperse assemblies, impose curfews and collective punishment, increase powers of arrest and search without warrant, and enforce severer penalties for the unlawful possession of arms. The peaceful population in occupied enemy country is protected by international law and if it acts against the occupying power it is now assured of a regular trial. Collective punishment for acts of individuals for which the population is not collectively responsible is forbidden; Article 39 of the 1949 Geneva Convention forbids such punishment even as reprisals. Killing of hostages was held to be a war crime by the International Military Tribunal at Nuremberg, but other Nuremberg judgments considered them permissible in certain circumstances. Article 34 of the 1949 Convention now forbids even the taking of hostages. It remains to be seen whether the new provisions will have much effect in practice.

The treatment of the population aims at preventing the revolt from spreading, restoring law and order and, in the country's own territory at least, winning over the population.

The theory is simple. As the German Manual puts it:

'the attitude of the population is of great importance in the fight against bands. Bands cannot continue in existence for any length of time in the midst of a population which entertains good relations with us. . . . The administration must see to it, by just treatment, planned and energetic government, and thorough and purposeful enlightenment, that the population is brought into the right relation to ourselves.'

The praxis, however, is complicated. From the anti-guerilla's point of view, if he shoots too early he may kill an innocent civilian, if he hesitates he may get killed himself. Seen from the population's angle the dilemma is equally inescapable:

'The people most severely affected are the villagers. At night a Viet-minh patrol makes them dig trenches across the road; in the morning a French patrol makes them shovel the earth back. If the Viet-minh suspect them of collaborating with the French, their village gets burned down.'[1]

It almost looks as if the anti-guerillas cannot do justice to their own troops without being unjust to the population, and vice versa.

If the population assists the partisans in the use of terror, the anti-partisans may be tempted to apply counter-terror. This was the method chosen at first by the German High Command of the Armed Forces. 'The troops available for securing the conquered eastern areas will, in view of the great extent of these areas, be sufficient only if the occupying power meets all resistance, not by legally punishing the guilty, but by spreading that kind of terror which is the only means of taking from the population every desire for opposition', reads one of the 1941 orders, or, as an army order, also of 1941, puts it, 'the population must be more frightened of our reprisals than of the partisans'.

It has been pointed out that counter-terror is only of value if over-

[1] London *Times*, February 15, 1950.

TREATMENT OF GUERILLAS AND POPULATION

whelming force is also available;[1] in that case, counter-terror is not always as useless as is frequently supposed. It stopped Mihailovitch in his tracks. On the other hand, it swelled the ranks of Tito's supporters. This may at first sight seem surprising, since the Germans scrupulously ascertained after each incident whether it was to be ascribed to Tito's or Mihailovitch's forces and then shot 'hostages of the group corresponding to that to which the culprit belongs'. The fact is, of course, that to Tito massive reprisals were welcome, as we pointed out in Chapter 1, while Mihailovitch was always mindful of the sacrifices of his supporters. It is therefore important for anti-guerillas who contemplate the use of counter-terror methods to assess carefully in advance the effect on the guerilla movement. They will also have to bear in mind that if they apply counter-terror the population will no longer give them any information.

In principle the use of counter-terror is to be deprecated. As Field-Marshal Lord Montgomery laid down in a directive in January 1947 in connection with the Palestine troubles, 'there can of course be no question of taking reprisals which would merely bear hardly on innocent people',[2] and the Germans too, in the face of mounting guerilla activities, were quick to mend their ways. By the autumn of 1942, they had issued orders to the effect that the population must not be put in a position where it was threatened by both sides, and that meant that reprisals should be taken against the people only if they voluntarily helped the partisans.[3] The same applied to collective punitive actions, and they were to be resorted to only if absolutely necessary; in this case the reasons had to be carefully explained to the people.[4]

There are two types of measures for the treatment of the population, the purely defensive and the positive. The first range from registration and search to collective fines and hostages, the second from resettlement to enlistment in the active fight against the partisans.

The anti-guerillas may find it necessary to register the population, issue identity cards, take a census of the resident population, screen non-residents and compel people to report at certain intervals. They

[1] Cf. Major D. M. R. Esson, 'The Secret Weapon-Terrorism', *The Army Quarterly*, 1959, p. 179.
[2] Cf. Field-Marshal Viscount Montgomery, *Memoirs*, London 1958, p. 469.
[3] Major Edgar M. Howell, *op. cit.*, p. 120.
[4] *Ibid*, p. 92.

may mount search operations; a cordon is laid for this purpose and everybody within the cordon must pass before a team of police interrogators furnished with black lists. The police are, if possible, assisted by ex-terrorists who, hidden behind a screen, pick out terrorists and partisans, usually in return for a free pardon.[1]

While guerilla helpers are punished, doubtful elements are isolated, by way of curfew, detention, or evacuation. An outbreak of terrorism is countered by a curfew, usually for all inhabitants during the night and for certain or all male age-groups in day-time, preventing the people from keeping contact with terrorists or carrying out acts of terrorism themselves. The curfew, coupled with food rationing, also serves to prevent food from reaching the terrorists. Suspects are detained, and it is usual to keep their cases under constant review. If these measures remain unsuccessful and partisan helpers are found again and again in a village, the entire village may be evacuated. This regrouping may or may not be effective. As Major O'Ballance points out, one-and-a-half million Muslims from selected areas in Algeria have been moved into camps without any attendant change in the war situation,[2] while on the other hand, according to Major Howell, captured Soviet partisan documents show that the evacuation of the population in partisan areas, coupled with the destruction of all farms, villages etc., was successful.[3] The destruction of villages may have done harm to the partisans but is seems to have harmed the Germans even more, because they laid down in 1944 that collective measures against the population of an entire village, including the burning down of houses, are only permissible in exceptional cases, and then only by order of a divisional commander.

No measure is more self-defeating than collective punishment. It is bound to alienate innocent people.[4] Worse, it is futile: the innocent can do nothing to escape punishment and the guilty, that is to say those who have information on the partisans, have the choice of losing either their property by order of the anti-guerillas if they do not tell, or their life by order of the partisans if they do.

---

[1] Cf. Major B. I. S. Gourlay, 'Terror in Cyprus', *Marine Corps Gazette*, September 1959, p. 33.
[2] Major Edgar O'Ballance, 'The Algerian Struggle', *The Army Quarterly*, October 1960, p. 97.
[3] *Op. cit.*, p. 93.
[4] Cf. Brian Crozier, *op. cit.*, p. 195.

TREATMENT OF GUERILLAS AND POPULATION

The same consideration applies to the taking of hostages. Major Esson has suggested that hostages should only be taken in order to compel the inhabitants to give information required for the protection of the lives of one's own troops. But a further safeguard seems required, and this applies to all forms of collective punishment: it should only be permissible if the troops can protect the inhabitants from partisan revenge.[1] Only then can the people make a moral choice and only then can collective measures against them be morally justified. Only then can it be evident that the anti-guerillas are doing justice not only to themselves but also to the population. This safeguard must be present, bearing in mind that the measures still hurt innocent people.

Effective protection against guerillas is also a basic requirement for success in reforming the people. If the population is willing, it will provide its own protection in the form of militia and home guards, and this local self-defence system needs buttressing by centrally located more regular forces which can quickly give assistance when required. Sometimes resettlement will be necessary before the people can be induced to provide self-help. Sometimes self-help will not be available on any terms and resettlement will then be required to reduce the defence burden, which will fall on the anti-guerillas alone. In all these cases resettlement is not a punitive measure applied in retaliation for some crime committed by the partisans or a warning for the future, but a purely protective measure to safeguard the people against partisan attacks and to withdraw them from partisan influence.

The concentration of the population into new villages can be compulsory, as it was in Malaya and Kenya. In both countries the population was spread in the smallest possible units over large areas. We have already mentioned the Malayan resettlement scheme. There, and in Kenya too, the new villages provided for the inhabitants a better life than they had had before. In Kenya schools, churches, first aid centres and sports fields were built, and the villages were protected by Home Guards formed under loyalists in the Kikuyu land unit. There was a certain risk in arming and training men who might not be reliable, but it was taken by General Erskine. As he has said, 'I decided to give the Home Guard maximum support and I never regretted it.' They manned well-sited and well-defended strong

---

[1] This seems a better criterion than the above-mentioned German one, viz. if the population does not voluntarily help the partisans.

points and were reinforced by former Mau Mau. They had European leadership and a regular full colonel acted as adviser. 'It would have been impossible', is General Erskine's verdict, 'to have achieved villages without the Home Guard.'[1]

Resettlement can also proceed on a purely voluntary basis and such a scheme was adopted in Cambodia during the Indo-Chinese war. As long as the government there tried to pacify the country by a system of defence posts and commando units, the population refused to co-operate; it did not give the information on which the commandos were supposed to act. Here too the population was widely dispersed and lived in fear of the rebels; by day they were not visible, by night they could exercise persuasion and terror.

When it was first announced that those who wished could move to new villages, more than half the people responded at once and more followed later. Here too the new villages provided some amenities, such as water, schools etc. It was ensured that the villagers should be able to follow their previous occupations. And finally, everybody took part in defence. Each village, with army help, was transformed into a strong point; several villages were grouped together into a zone. A mobile counter-guerilla commando was based in the centre of the zone. The civil administration and the future community leaders received civic and military instruction in courses lasting about three weeks and the government gave economic help to the new villages. The scheme has proved valuable.[2]

Some resettlement also took place in the Red River delta but apparently without much success; the militia was not reliable.

It has already been mentioned that the village Home Guard in Kenya proved trustworthy, and the same applies to the Kenya urban Home Guard and the Forest and Farm Home Guards. In Malaya also it earned nothing but praise, and it produced there a sense of self-reliance, comradeship and community spirit.[3] Why, then, did the French experiment in Tongking fail? As Brian Crozier puts it, 'a promise of real independence undiminished by reservations in favour of France, and implemented in good faith within the limits imposed by the state of emergency, would have given Bao Dai the power he needed and a cause to fight and die for. But the French

---

[1] *Loc. cit.*, pp. 15 and 16.
[2] Capitaine André Souyris, *loc. cit.*, pp. 686 f. is the authority for the above facts.
[3] See Federation of Malaya, *Annual Report 1955*, p. 431.

acted in bad faith. They never intended to give ... real independence to the country.'[1] A Frenchman, Jean Farran, sees the problem somewhat differently: the French changed their war aims; at first the war was fought to maintain French domination and then, after independence had been granted, in order to fight communism, but this could not convincingly be demonstrated because the French were left to fight it out alone. 'In any case, if we (French) hardly knew what we were fighting for, the millions of Vietnamese knew it even less.'[2] Whichever interpretation the reader prefers, this much is evident: people will fight only if they have a cause and know it.

It is quite obvious that the same rule also applies if the anti-guerillas want to win the population over to their cause: the population must know these aspirations and share them. There are three ways to achieve this conversion; by education (propaganda), by promises, and by example.

The task of propaganda, then, is to explain to the people why the war is being fought and why they have a part in it. The government forces in the Philippines which in the post-war years achieved victory over the communist-led guerillas, the Hukbalahap, used to hold after each screening operation a rally consisting of short educational talks by the civil affairs unit on citizenship, democracy, communism and the role of the army in the anti-guerilla fight. 'These psychological efforts did much to counteract (communist) efforts along similar lines.'[3] The French, too, used search operations in Indo-China as an opportunity for addressing the people, telling them how damaging their passivity and silence were to their own interests.[4] These two approaches mark the propaganda range, from the sublime to self-interest. But propaganda about citizenship and democracy, without granting them, would be just as futile as the appeal to self-interest when the population is at the mercy of the guerillas. Propaganda must be backed by reality. It must be informative. It must take people into its confidence. Because these conditions were fulfilled in Malaya, the State and Settlement Information officers and the District War Advisory Committees succeeded; it was certainly helpful that the Committees were composed of local inhabitants, and that a second

[1] *Op. cit.*, p. 116.
[2] Jean Farran, 'La leçon de Dien Bien Phu', *Bulletin Militaire*, August 1956, p. 583.
[3] Cf. for the above Lt.-Colonel Luis A. Villa-Real, 'Huk Hunting', *Army Combat Forces Journal*, November 1954, p. 35.
[4] Cf. *Tropique*, January 1954, p. 52.

line of communication to the people existed through the elected council members at village, local, municipal and State level. The people, briefed in this way on the progress of the war in their area and the measures to be taken against the terrorists, responded by showing increased confidence in the government's victory and their optimism could not fail to affect the guerillas' morale.[1]

In Malaya psychological warfare against terrorists and emergency propaganda among the public was conducted at top level by different agencies, the former by the Director of Operations' Staff, the latter by the Director of Information Services, but these worked in close contact. In the field they had common agents in the State and Settlement Information Officers already mentioned.

Propaganda is to some extent interwoven with promises. The French at one time put great trust in this weapon in Algeria and their Psychological Warfare Department was very active there from 1957 on, but with little effect. The London *Times* found that 'much of this is rather heavy-handed, and in any case, what conviction can be carried by political assurances on the future, in the face of the legacy of broken promises in the past'.[2] It appears that this harsh appraisal was not entirely unjustified; the Psychological Warfare Department was dissolved in 1960.

It seems platitudinous to say that promises are only effective if people believe in them. They will do so if the promises, or at least some of them, are already implemented during the emergency. In Malaya the new constitutional structure was laid during the emergency, elections took place during the emergency, and there and in Kenya land grants were made during the emergency.

The troops, too, can play a vital part in winning the people's confidence. They work frequently under a disadvantage: they differ from the population in race, religion and customs, and they are foreigners fighting the people's own countrymen. But these differences can to some extent be bridged by the troops' keenness on the job and by their willingness to co-operate with the population. Again Malaya affords an example; units were affiliated to villages. As Brigadier K. R. Brazier-Creagh has pointed out,

'these units play a very full part in their villages. They sit on the

[1] Cf. Federation of Malaya, *Annual Report 1956*, p. 441.
[2] Cf. London *Times*, March 4, 1958.

TREATMENT OF GUERILLAS AND POPULATION

village committees, run games, play basketball, give demonstrations, give outings to the local inhabitants, organize band concerts, help in levelling sport fields, and in fact try to identify themselves with the life of their affiliated communities.'

Army doctors treated the local people.

'The army, with its resources, can help in solving some of the immediate problems and show that it is wholeheartedly behind the government in its effort to provide the improvements so much desired by the people.'[1]

As Brigadier Brazier-Creagh also points out, British troops in Malaya have given assistance in training the Home Guards. The Home Guards too can greatly contribute towards converting the people. They show them that there is no united front of the people's liberators against the government and that the government is not fighting against liberty, but against liberty on the terrorists' terms. They embody the spirit of self-reliance and by carrying the problems of distant campongs to the centre they forge a new link between government and people.[2] They will fulfil their task best if they are true volunteers.[3]

We have tried to show in the foregoing that severity against the guerillas and the population does not pay. It may be asked whether leniency does not create its own dangers. It does up to a point, if the terrorists can freely put pressure on the population, which may then yield too easily because it feels safe from anti-guerila reprisals. But we have seen that the Germans had learned that even then it is still wiser to treat the people with restraint. At any rate, the right answer to the problem seems to be to punish the partisans and not the population. Finally, can justice be reconciled with one's own security? Not always: if one terrorist hides in a village it will be safest to treat the entire population as terrorists; and where reprisal actions have to be taken, many innocent people will suffer for the misdeeds of the few. But it must be recognized that the dilemma need arise only where the lives of one's own troops are concerned and in this case

[1] Lecture by Brigadier K. R. Brazier-Creagh, 'Malaya', reprinted in *Royal United Service Institution Journal*, 1954, pp. 175 f.
[2] The last point is made in Federation of Malaya, *Annual Report 1955*, p. 431.
[3] As Commandant J. Hogard points out in 'Le soldat dans la guerre révolutionnaire', *Revue de Défense Nationale*, February 1957.

the security of one's own troops takes precedence over everything else; in all other cases security and justice can be reconciled, and will be reconciled if it is realized that the fight is not only *against* partisan revolt but also *for* the people's support.

CHAPTER 11

ON WHOSE SIDE VICTORY?

Literature on guerilla warfare does not abound in comparative studies—one might put it more crudely by saying that there is in fact not a single one. For this reason it will hardly be found possible to develop authoritative general theories on guerilla war. It is therefore all the more welcome that one attempt at formulating such a doctrine has been made in one sector of guerilla war by a group of French officers who have analysed the Revolutionary War.

The French use the term 'revolutionary war' to indicate 'any armed combat undertaken by a minority which progressively controls the population and which supplies it with motives to act against established authority or an authority which it refuses to recognize'.[1] It may be doubted whether the East Berlin rising falls into this category; the theory itself is not concerned with risings against Bolshevism.

According to this doctrine the revolutionaries must first conquer the population (non-violent phase) before they can embark on war (violent phase). One of the exponents of this doctrine, Commandant J. Hogard, holds that there are five stages in the process:[2]

1. Propaganda cadres visit a future base, explore grievances of the population and exploit them in order to spread discontent.

2. The people are then organized politically into groups which supervise each other and are in turn supervised by committees controlled by activists. Opposition to the government is worked up all over the country through propagandists, agitators, ideologists and politicians, traitors are exposed, terror is brought into play and strikes and riots take place.

3. Guerillas now engage in small-scale actions and propaganda and sabotage activities increase.

[1] Ximenes, 'La Guerre révolutionnaire et ses données fondamentales', *Revue Militaire d'Information*, February/March 1957, p. 11.
[2] *Guerre révolutionnaire et pacification*, ibid, January 1957, p. 11.

4. As soon as the government withdraws from exposed areas, the revolutionaries create liberated zones there. The revolutionaries may now set up a government. A regular army begins to be formed and raids and terrorism are extended to government areas.

5. Finally the revolutionaries go over to the military and psychological offensive.[1]

This classification is obviously modelled on the methods used by Ho Chi-minh in Indo-China, but, as we have seen in Chapter 3, Mao Tse-tung proceeded quite differently when he formed new bases: he took them by force and reformed the population afterwards. What is more, the description does not apply to Yugoslavia, because Tito already had popular support in many parts of the country when he needed it; it does not apply to Kenya and Cyprus because the rebels did not intend to defeat the British forces in the field; and it does not fit Castro's war in Cuba. The Palestine revolt and the Indonesian war, both designed to bring foreign public pressure to bear on the occupying powers, are also not covered by this classification. Finally, as Mao said, there is little or no possibility of forming base areas in small countries, and the above scheme could not therefore apply to them either.

Another classification has been introduced by Ximenes, who differentiates between crystallization, organization and militarization, which are all linked together.[2] Thus if a cadre of activists has imbued people with a sufficiently firm conviction (crystallization), it can charge this group with certain responsibilities (organization) and then proceed to carry out a small raid on a police post (militarization). Or a military success (militarization) is exploited by propaganda, leading to crystallization, on the basis of which a village council is formed (organization), or again, when village, district and provincial councils have been formed in a new area (organization), propaganda and courses in political education result in volunteers coming forward (crystallization) who then go into combat (militarization).

Ximenes sees one practical application of this classification in the following: if the crystallization is good but organization and militari-

---

[1] For a fuller description of the five stages cf. Peter Paret, 'The French Army and la Guerre révolutionnaire', in *Royal United Service Institution Journal*, vol. 104, 1959, p. 60.

[2] Ximenes, 'La Guerre révolutionnaire et ses données fondamentales' *loc. cit.*, pp. 16 seq.

zation are weak, a sufficiently dense and skilful military occupation by the government forces will restore the situation.

Now this was exactly the position facing Britain in Cyprus: crystallization good, organization and militarization weak. British forces did occupy Cyprus. But the presence of troops did nothing by itself to turn the tables. It was the constant patrolling, the sweeps and the hunts, and the efficient employment of helicopters which changed the situation. Here again, the theory does not seem to be confirmed in practice.[1]

We have tried to show in Chapter 3 when bases are necessary and when they are not. We have also tried to explain the function of the base and in particular its strategic importance in certain circumstances. We have finally tried in that chapter to indicate when the insurgents require semi-regular forces and when they do not. If we have succeeded, it follows that there is no justification for depicting the base as the concomitant of all revolutionary movements. And it will be remembered that in Tito's fight it did not always seem to matter how many village, district and provincial councils (organization) had been formed by his movement; in his case crystallization and militarization owed little or nothing to the interplay of organization. Not all revolutionary movements can therefore be assessed according to the quality of the organization.

Perhaps this book has also succeeded in showing that while some general rules for partisan warfare can be devised, a general theory for revolutionary movements cannot be developed because, sometimes at least, each goes its own individual way. The foregoing analysis seems to reinforce this view.

However, this criticism should not detract from the positive contributions of the doctrine. It rightly stresses the need for careful study of 'the cause, the efficacy and the fundamental facts of the revolutionary war'.[2] The first requirement, as Capitaine Souyris points out, is a proper appreciation of the situation; one must not mistake a revolutionary war for just another 'traditional' rising. This mistake was made in Algeria in 1954-55 where it was at first believed that the

---

[1] Quite apart from the difficulty of correctly assessing abstract conceptions in terms of good or weak. Furthermore, one can see that if crystallization is good in half the country it can fairly be called 'good'; but can it still be described as 'good' if it is good only in a quarter of the country or in the hills and villages but not in the cities and towns?

[2] Capitaine A. Souyris, 'Les conditions de la parade et de la riposte à la Guerre révolutionnaire', *Revue Militaire d'Information*, February/March 1957, p. 94.

outbreak was the usual tribal insurrection which could be put down with small forces and that the police could subsequently maintain order, a mistake not dissimilar to that made in Kenya where the existence of an emergency and its nature were not recognized in time. The theory also stresses that one cannot hope to end a revolutionary war by remedying the grievances of the population exploited by the revolutionaries; such terms are unacceptable to the revolutionaries because they are not fighting for this end; they only want to attain power. In fact, the French found that, if reforms were instituted, resistance only became stronger.

Nor, in the French view, can the war against the revolutionaries be won by military means alone. As Commandant Hogard points out, the revolutionaries, beaten in stage three, four or five of their development, can simply start again with stage one or two.[1] The revolutionaries' infrastructure, their politico-military network, must be destroyed to bring the war to an end.

This, then, is the essence of the French doctrine and, as the reader will remember, Colonel Woodhouse has expressed it earlier in these words: 'The art of defeating the guerillas is . . . the art of turning the populace against them.'[2] The only question is how this can best be achieved.

The doctrine differentiates here between the two phases of the revolutionary war, the non-violent and the violent. In the pre-violent phase the government must act in all spheres—political, economic, social and psychological—if it wants to keep the confidence of the masses, and it must maintain constant touch with them. Apart from reforms and public guidance, police and court action are required, and the army must control the danger zones by setting up small posts and mobile units for quick intervention in emergencies. The army must not only control the area but also supervise and guide the inhabitants. Furthermore, a good intelligence service is required. On this basis it should be possible to fight in case of an outbreak—the second phase of the revolutionary war—on the opponent's 'terrain' which is the population. The population must be mobilized, physically—into self-defence units—and morally, by the lawful authorities. The army must protect communications and installations. It must pacify the country by means of a network of small posts which must be spread and augmented, and it has the task of winning over

[1] *Loc. cit.*, p. 13.　　　　　　　　　　　　[2] Cf. above, Chapter 2.

the population. Finally, the army must destroy the enemy's forces. But this can, in the French view, 'never be an end in itself; it is only useful insofar as the dismantling of the rebel infrastructure is speeded up. The army's measures must therefore be guided as much by political and psychological considerations as by military ones—perhaps more so.'[1] The government must resist the temptation to rely exclusively on its security forces and to believe that it could annihilate the rebels by military means.[2] But the population cannot be won back unless the armed forces are very densely spread over the country in order to reduce the rebels to impotence.[3]

To some critics the fundamental thesis that the army 'is not only a fighting force but an agency of education and moral regeneration'[4] may go too far. There are other objections. Is it really always practicable to keep the country 'very densely' occupied? Keeping her other commitments in mind, could France really have done it in a country the size of Indo-China? Secondly, is it really true that 'the army certainly plays an important role in the violent phase but contrary to what one might believe its action is not decisive'?[5] Was it not decisive in ending the Greek civil war? The government there, it is true, gradually won the support of the population, but what ended the war was the successful army attack on the vital bases; otherwise, with Albanian support, the guerillas could have carried on the war for many more years and perhaps even broken the country. And again, was not the army decisive in ending the Kenya emergency? Is it not also correct to say that nothing is more likely to win the support of the population than military success against the rebels? To win this support can take a long time; to restrict army action on account of political and psychological considerations may give the rebels the time needed to strengthen their semi-regular forces and help them to gain victory in the field.

There is one final consideration. The doctrine concentrates so much on the population and how to win it over that it seems to have overlooked the fact that rebels can be won over too. Backed by military successes, the rewards system in Malaya finally cleared the jungle of guerillas, and again, backed by military successes, the amnesty in Kenya induced almost 1,000 Mau Mau to surrender.

[1] Summing-up by Peter Paret, *loc. cit.*, p. 64.
[2] Capitaine A. Souyris, *loc. cit.*, p. 107.   [3] Capitaine A. Souyris, *loc. cit.*, p. 108.
[4] Peter Paret, *loc. cit.*, p. 68.   [5] Capitaine A. Souyris, *loc. cit.*, p. 110.

PARTISAN WARFARE

The wars were thus considerably shortened. The rebels came out of their hide-outs because they had been the object of some 'decisive' actions by the government forces: the guerillas knew that they had been beaten. The revolutionaries' strength may not rest in their fighting forces, but once they are annihilated one can quickly round up the politico-military leaders and smash the infrastructure from the top.

In summing up we would say this: it appears that in its later stages the war in Malaya was conducted with the right mixture of military, political, economic, social and psychological ingredients. To change the emphasis in the way the French doctrine proposes to do, does not seem warranted in the light of British experience and success.

Anti-guerilla wars can be fought successfully the way they have been fought—vide Greece, Malaya, Korea, the Philippines, Kenya, and Cyprus—and there is hardly any justification for quarrelling with the concept of these operations. But there are two aspects, common to all guerilla wars, which might warrant a new approach. We have touched on both of them before: the inordinate duration of these wars, and the excessive superiority in numbers required of the anti-guerillas.

The guerillas, as we have also said before, want a protracted war because it wears down the enemy and gives them time to transform their inferior forces into superior ones. Mao Tse-tung has explained this in his series of lectures *On the Protracted War*, given in 1938 to the Association for the Study of the Anti-Japanese War in Yenan:

'The enemy is strong and we are weak, so we face the danger of subjugation. But in other respects the enemy has shortcomings and we have advantages. The enemy's advantages can be reduced and his shortcomings aggravated by our efforts. On the other hand, our advantages can be enhanced, and our shortcomings remedied, by our efforts. Therefore, we can win our final victory and avert subjugation, while the enemy, ultimately doomed to defeat, cannot avert the collapse of his whole imperialist system.'[1]

'The advocates of the theory of a quick victory do not understand that war is a contest of strength, that it is vain for them to wish to

[1] *Selected Works of Mao Tse-tung*, vol. ii. *op. cit.*, p. 181.

wage strategically decisive battles to hasten towards the path of liberation before definite changes in the relative strength of the warring parties have taken place.'[1] 'The fact is that the disparity in strength between the enemy and ourselves is at present so great that the enemy's shortcomings have not been and cannot yet be developed to the required degree to offset his strength, while our advantages have not been and cannot yet be developed to the required degree to compensate for our weakness; therefore there cannot yet be parity but only disparity.'[2]

To change the balance in China's favour, the protracted war must pass through the well-known three stages:

'The first stage is one of the enemy's strategic offensive and our strategic defensive. The second stage is one of the enemy's strategic defensive and our preparation for the counter-offensive. The third stage is one of our strategic counter-offensive and the enemy's strategic retreat.'[3]

And this model exposition has been faithfully copied ever since.

The second concomitant of guerilla war, the extraordinary demands on the anti-guerillas' manpower, finds its explanation to some extent in their static protection and guard duties. But the actual figures give food for thought. To give only two examples: in Greece the Army had a tenfold numerical superiority over its adversary, in Algeria 30,000 trained FLN guerillas are opposed by half a million French soldiers, that is almost one Army brigade for each guerilla company.

Now, as Colonel Woodhouse has pointed out in the statement already quoted,[4] the outcome of a guerilla war depends on the morale of the people and the availability of good communications. It takes a long time to convert the population, it takes time to build roads. But must both processes take so long that they present the guerillas with the opportunity for a prolonged war, and must the army employ so many men on static duties?

Both problems are peculiar to guerilla war. In orthodox wars there is no battle for the mind of the people, and an army, facing a seven-

[1] *Selected Works of Mao Tse-tung*, pp. 192/3. [2] *Ibid*, p. 181.
[3] *Ibid*, pp. 183/4. [4] Cf. above, Chapter 2.

teenfold superiority, could not fight a prolonged war. In orthodox war the compression of a campaign into the shortest possible space of time may be of strategic importance; the Blitzkrieg is an example. But in anti-guerilla war it ought to be, at all times, the supreme and dominating concept, in order to prevent the guerillas from turning weakness into strength.

Where the population is loyal, the number of troops deployed for static protection duties will be limited, because there will be enough volunteers for the Home Guards, and if the population is widely spread it can be concentrated in new villages, requiring fewer guards. The anti-guerilla war, supported by the people, ought to be of relatively short duration.

The problem, then, is whether the duration of anti-guerilla wars can be shortened if the population is hostile or indifferent. This was, in fact, the state of mind of the people in Cyprus, certainly in the villages and, as far as the less well-to-do Greeks were concerned, also in the towns. And yet, the British were about to win that war. It was probably the first time in the history of guerilla war that victory for the anti-guerillas was around the corner in spite of the unhelpful attitude of the population.

How could this come about? Major Gourlay gives the clue by pointing out that helicopters made it possible to get at the enemy before information about an impending operation reached him.[1] A helicopter carrying troops travels faster than the warning given by members of the population to the guerillas.

The intervention of the helicopter makes the attitude of the people appear less important than both sides have hitherto rightly assumed. For the guerillas the population will still remain a source of manpower and supplies but no longer of information. The anti-guerillas have to some extent shown themselves able in the past to stop food supplies to the guerillas, by rationing the population and by curfews, but they were never able to prevent the people from giving information to the guerillas. If guerillas can now no longer be forewarned and suffer bigger and more frequent losses as a result, the people will also become less eager to volunteer as guerillas. Under the new conditions, exemplified in Cyprus, a hostile or indifferent population is no longer the menace it used to be.

The advent of the helicopter does not, of course, solve all prob-

[1] Major B. I. S. Gourlay, *Terror in Cyprus*, loc. cit., p. 34.

ON WHOSE SIDE VICTORY?

lems. It reduces the dangers to the anti-guerillas from a hostile population and it reduces them considerably. It diminishes the value which popular support has for the guerillas. But even so, this is still some way short of perfection, because what the anti-guerillas need is a population friendly to them and willing to give all the information it has. Yet even here the course of events in Cyprus shows that the necessary intelligence can be obtained by other means, especially by the troops themselves.

Where plenty of helicopters are available, even the need for good road communications fades into the background. But not entirely; because the transport of heavy equipment still depends on roads, transport by helicopters is expensive and they cannot be regarded as maids-of-all-work.

The lesson seems to be that in an anti-guerilla war in which helicopters are used in sufficient numbers, a neutral or even hostile population can do little to help the guerillas in protracting the war if the anti-guerillas have other sources of information available.

What other sources are there? Loyal members of the population, Home Guards, *Jagdkommandos*, reconnaissance patrols, prisoners, and above all, pseudo-gangs. Results in Kenya with pseudo-gangs speak for themselves and there is no need to recall what has been said before in this book. If pseudo-gangs can possibly be formed, they must be formed, in every anti-guerilla war.

If the commander of the anti-guerilla forces has both helicopters and pseudo-gangs at his disposal or, failing the latter, uses his other sources of information well, the road to victory is considerably shortened because the conversion of the population is no longer absolutely essential to success. Also no more anti-guerillas need be deployed in hostile areas than necessary to stop food supplies to the guerillas.

That the anti-guerillas now have the means of shortening the protracted war and denying the guerillas time to achieve superiority is by far the most important development in modern anti-guerilla warfare.

The guerillas have not yet found the answer.

It is unlikely that they will.

This was shown in Algeria. At first French helicopters would land in the immediate vicinity of the enemy. When the rebels had learnt how to attack helicopters the French developed new tactics, providing for maximum troop debarkation when the terrain permitted and the selection of alternative landing zones when the rebel fire was prohibitive in the primary zone.[1] The anti-guerillas often have the advantage of complete mastery of the air and even where guerillas are fighting in association with their own Army its air-force can give them only very limited assistance against anti-guerilla surprise landings.

We have tried to show in this book when the military machine is superior to guerilla opposition and under what conditions the guerillas have the advantage. It remains to summarize the strategic and tactical principles of guerilla warfare which we have discussed in this book.

Mao Tse-tung has outlined the strategy of guerilla warfare as follows:

'1. On our own initiative, with flexibility and according to plan, carry out offensives in a defensive war, battles of quick decision in a protracted war, and exterior-line operations within interior-line operations;

2. Co-ordinate with regular warfare;
3. Establish base areas;
4. Undertake strategic defensive and strategic offensive;
5. Develop mobile warfare; and
6. Establish correct relationship of commands.'[2]

Mao's famous Ten Military Principles codify guerilla tactics as follows:

1. Attack first scattered and isolated enemies, later attack strong enemy forces;

2. Take first the small towns; later take the large towns;

3. The major objective is not to hold cities but to annihilate the enemy's fighting strength;

4. Concentrate absolutely superior forces for every battle;

5. Do not fight unprepared engagements or engagements in which victory is not assured;

6. Fear neither sacrifice nor fatigue. Train to fight successive engagements within a short period;

[1] Cf. Major David Riley, *loc. cit.*, p. 25.
[2] *Strategic Problems in the Anti-Japanese Guerilla War*, pp. 122/3.

7. Destroy the enemy while he is moving;
8. Take first weakly defended cities; when conditions are favourable take those with medium defences. Wait until our advantages are enhanced before you take strongly defended cities;
9. Capture arms from the enemy so as to arm yourself;
10. Use intervals between two campaigns skilfully for resting, regrouping and training troops . . . but do not let the enemy have breathing space.[1]

For anti-guerilla warfare, which we consider now, there is one basic rule: the anti-guerillas must not slavishly apply orthodox strategy and tactics or rush in the big battalions in forest and jungle searches for small guerilla bands; the enemy does not as a rule intend to hold ground and will be gone before they have a chance to find him. The anti-guerillas must learn the special rules of guerilla and anti-guerilla warfare. The present study has tried to bring out some of the rules of anti-guerilla warfare and they may be summarized as follows:

1. Fight a short war. Do not give the enemy time to gain strength and superiority.
2. Attack first the enemy's strongest points. Go for his bases.
3. The major objective is not to hold a front line or a defence line of forts but to annihilate the enemy's fighting strength.
4. Do not hold more ground than you can afford to. If your forces are insufficient to seal up the enemy, establish base areas.
5. Do not give the enemy the chance to encircle you.
6. Keep up the offensive spirit among your troops.
7. Surprise is a main element of successful anti-guerilla tactics. Use your air force to achieve surprise and to increase the profitable deployment of your troops.
8. Penetrate the enemy by forming pseudo-gangs.
9. Isolate the enemy by denying him access to the population and cutting his supply lines.
10. Improve your communications.
11. Keep your static defences to a minimum. The friendly population will provide Home Guards for self-defence. In hostile areas

[1] Cf. Mao Tse-tung's *Speech to the Central Committee of the Chinese Communist Party*, of December 25, 1947.

guard troops are only required in order to prevent supplies from reaching the enemy. If necessary the population must be resettled.

12. Do not treat the population and guerilla prisoners harshly. Reprisals should be taken against the population only if it is protected but still helps the partisans.

Wars are not won by rules and guerilla wars are no exception. Besides, the enemy has his rules too, and only one side can win. But rules can help a commander to come to the right decision. On the few occasions when General Giap erred he had neglected Mao's maxims, and the German anti-guerillas would have been better served, had their Manual come out earlier.

Though the aims of anti-guerilla warfare are clear, their realization depends on local conditions. But precedents are an invaluable guide. The anti-guerilla commander must feel his way and obtain the best possible help: the humble man who has local knowledge, the counter-gangs who acquire it first-hand, the imaginative army intelligence or police officer and the even more imaginative Staff officer who, knowing how others did the job before them, improve and adapt tested techniques and apply new ones.

The best brains have developed guerilla warfare until it has become the serious menace it now is. It has also become a continuous menace: since 1927, without a break, some guerilla war has been fought somewhere, and frequently there has been more than one guerilla war going on at a time. But guerillas are not invincible and, let us repeat, Malaya, Kenya, Cyprus, Greece, the Philippines, and last but not least, Korea, are eloquent proof.

Guerilla warfare has long been treated in the East as a proper subject for military science, and it is time that the same service were done for anti-guerilla warfare.

Military science cannot prevent the outbreak of future guerilla wars, but it can perhaps assist the military in assembling case material, drawing conclusions, however tentative, and submitting views on how to shorten the duration and limit the sacrifices of all concerned.

This, at any rate, is the purpose of the present study.

CHAPTER 12: APPENDIX

ON PARTISAN WARFARE IN WORLD WAR II

Partisan warfare played only a minor part in World War I. In World War II partisan movements sprang up in almost every country invaded by Germany and Japan.

This marked difference in the attitude of the population towards the conqueror may seem surprising at first sight. There were, however, a number of reasons which accounted for this change and furthered widespread guerilla activities in the last war. The main factors may be summarized as follows:

1. There was, in the first place, the technical aspect. Never before had there been so many opportunities for partisan action. The dependence of modern armies on their lines of communication and the wide extension of the battlefields, far away from home bases, provided guerillas with an abundance of targets, while radio and aircraft facilitated contact and supply.[1] The interest which Great Britain took from 1940 onwards in special operations, and subsequently the Soviet exhortations to their own people to foment guerilla warfare everywhere, stimulated the desire to resist in countries overrun by the Axis and helped them to turn it into action. This leads to the second factor, the psychological one.

2. The very violence of the shock administered by the Blitzkrieg deepened the sense of humiliation in the defeated countries, especially where Quisling governments were ready to sell out to the conqueror. Since the war was an ideological conflict, active antagonism was reinforced by a clearer appreciation of the consequences of final defeat. In other words, the war aims were more readily understood than in World War I, and while the loser in World War I would have to pay compensation, cede some territory and put up with foreign

[1] Cf. F. O. Miksche, *Secret Forces, The Technique of Underground Movements*, London 1950, pp. 35 f.

occupation for a limited time, the vanquished in World War II, had the Axis been victorious, would have had to submit to the New Order which many were unwilling to accept. The course of the second World War gave the defeated a chance which the first one had never offered to anything like the same extent. In the first World War practically all regular armies on both sides, with the notable exception of Russia, fought on to the end and there was hardly any room for partisans. In the second World War the Blitzkrieg had taken its toll, but as long as Great Britain and later on her allies remained in the war, defeat for the conquered was not final. Here the example of the first partisan in Europe, Mihailovitch, had its effect, and subsequently Tito's successes fired the imagination; Yugoslavia showed the world that a country could withstand the invaders even after its army had been defeated. But what roused armed opposition more than anything else in World War II was the short-sighted German occupation policy.

3. The third factor which did not play any part in the first World War was political. Communists did not fight the Germans until they attacked Soviet Russia. On the contrary, communist efforts before that time were rather directed towards weakening the British and French war effort. The French communists in particular set out to undermine the morale of the population and the army while it was still in the war, because it was an 'imperialist' war.[1] But when the communists took to partisan warfare they not only mobilized large numbers of their supporters but also spread the spirit of resistance.

4. The fourth factor which made many join the partisans was simply fear, fear of the enemy or fear of the partisans, and this again was a phenomenon peculiar to the last war. Finally, forced recruitment also played some part.

The partisan movements, as a whole, never came under a unified command. When the battle front approached their area they received their orders from the theatre commander. But up to 1944 a certain over-all control of partisan warfare outside Soviet Russia lay in British hands.[2] This was established in various ways:

1. Britain dispatched agents to a number of countries in order to form sabotage groups composed of local people.

[1] For particulars cf. O. Heilbrunn, *The Soviet Secret Services, op. cit.*, pp. 56 seq.
[2] In March 1944 relations with the French Resistance came under U.S. authority.

APPENDIX

2. Britain gave the partisans the tools for their tasks. She could thus exercise her preference for one movement or another and make the partisans amenable to certain requests in order to qualify for aid themselves or prevent a rival movement from becoming its sole recipient.

3. By sending Military Missions to several countries, she could try to exercise her influence in an even more direct way.

4. Finally, Britain could send Special Forces into partisan territory to bolster up partisan morale.

Whether British influence was accepted in praxis did not depend on the political outlook of the various guerilla movements. The dividing line was in fact not between communist and non-communist partisans: the French and Italian communist movements executed British—and American—directives, but the Greek communists did so only intermittently, and the royalist Mihailovitch, from 1943 on, not at all. We will discuss the reasons for the communist line-up later on.

The British furthered resistance in every country where the Germans, Italians and Japanese were, while American interest was restricted to the future battle-zones—North Africa, the Philippines, Italy and France. When the British and American spheres coincided, as they did in Italy and France, it was not always easy to establish a common viewpoint, and even where American interest was less direct, or remote, as in Greece and Yugoslavia, American and British policy could not always be reconciled. The US continued to support Mihailovitch after Britain had withdrawn her support from him, and when Britain used force in Greece, in December 1944, in order to restrain the communist partisans, the US protested.[1] But in spite of these differences the effectiveness of partisan resistance was not affected.[2]

It would be wrong to say that American aims were purely military and British aims predominantly military. The American reluctance to accept General de Gaulle as the representative of Free France and

[1] See also Norman Kogan, *American Policies towards European Resistance Movements*. Paper read at the Second International Conference on the History of the Resistance, Milan March 26/29 1961.

[2] Although the withdrawal of British troops from Italy in December 1944 to put down the armed rising in Greece affected military operations in Italy. Colonel Woodhouse, *The Greek Resistance 1942-44*, loc. cit., p. 387.

the desire to build up General Giraud as a counter-force were influenced by purely political considerations. On the other hand Britain did not allow her political preferences for the post-war period to interfere with the military requirements of the moment.

As far as Britain was concerned, it was mainly in Greece that political and military interests became intermixed. It was not Britain's doing that politics came into the picture at all. In 1942 a Special Operations Executive brief stated that 'SOE interests lie in promoting unity inside Greece for the purpose of resisting occupying forces and stimulating rebellion, and we are not interested in post-war Greece'.[1] A Foreign Office directive of March 1943, laid greater stress on unity, but defined it as meaning not only unity between the resistance movements but also between them and the King and government; it also made it clear that, subject to special operational necessity, the British Liaison Mission—commanded by Colonel Woodhouse—should 'always veer in the direction of groups willing to support the King and government'.[2] Half-a-year later ELAS acquired most of the weapons of the Italians surrendering in Greece, thus attained military supremacy, attacked other guerilla forces in Greece in order to gain control of the country, and in the spring of 1944 set up a communist-controlled government which ruled over at least half the country.[3] Again, Britain made it clear that the choice of a post-war government was a Greek affair, but equally stressed that Greece could not find 'constitutional expression in particular sets of guerillas, in many cases indistinguishable from banditti who are masquerading as the saviours of their country while living on the local villagers'.[4] The value of guerilla operations, not on an important scale since the summer of 1943, had in the meanwhile become slight.[5] At the end of 1944, after Germany's withdrawal from Greece, British troops intervened to save her from mob rule.

In Yugoslavia events took a different course. Britain was less interested here than in Greece that 'constitutional expression' should be assured, the reason being that the monarchy was held to be a unifying force in Greece, as the plebiscite in 1946 confirmed, but

[1] Cf. Professor F. W. D. Deakin, *Great Britain and European Resistance*. Paper read at the Second International Conference, as above.
[2] For the full text cf. Colonel C. M. Woodhouse, *loc. cit.*, p. 377.
[3] Colonel C. M. Woodhouse, *op. cit.*, pp. 385/6.
[4] Sir Winston Churchill, *The Second World War*, vol. v, London 1952, pp. 481/2.
[5] Colonel C. M. Woodhouse, *loc. cit.*, p. 382.

APPENDIX

that in Yugoslavia Tito would be the people's choice. The British position is probably best summed up in a cable sent to the Foreign Office on December 25, 1943, by the British Ambassador to the Royal Yugoslav Government:

'Our policy must be based on three new factors: The Partisans will be the rulers of Yugoslavia. They are of such value to us militarily that we must back them to the full, subordinating political considerations to military. It is extremely doubtful whether we can any longer regard the monarchy as a unifying element in Yugoslavia.'[1]

Thus the differences between the liaison organizations in Yugoslavia and Greece become understandable: the British Liaison Missions to Yugoslavia were not charged with uniting the guerilla movements in the country or uniting them with the King and government. Furthermore, there was not one Mission as in Greece, dealing with all the partisan movements, including the communist partisans, but there were two, the Mission to Mihailovitch until December 1943, and from September 1943, the Mission accredited to Tito's forces. Finally,

'the directive issued to the British Mission to Tito laid particular emphasis on the purely military aspect of the situation . . . Politics were a secondary consideration.'[2]

Subsequently the political question appeared satisfactorily settled. On August 12, 1944, Sir Winston Churchill and Marshal Tito met in Italy.

'Tito assured me', Sir Winston wrote, 'that, as he had stated publicly, he had no desire to introduce the communist system into Yugoslavia. . . . The Russians had a mission with the partisans, but its members, far from expressing any idea of introducing the Soviet system into Yugoslavia, had spoken against it.'[3]

This leads to the next topic of our discussion, Soviet influence over

[1] Quoted in Sir Winston Churchill's *The Second World War*, vol. v, p. 414.
[2] Sir Fitzroy Maclean, *op. cit.*, p. 240.
[3] *The Second World War*, vol. vi, London 1954, p. 81.

the various partisan movements outside her own borders. It was restricted to communist movements. How and to what purpose was it exercised?

The general party line had been laid down as early as 1928. When the Congress of the Communist International met in Moscow that year, the thesis was advanced that a war waged by imperialist States must be transformed into a civil war of the proletariat against the bourgeoisie by means of revolutionary uprisings; circumstances permitting, and especially if the Soviet Union were attacked, communists must form guerilla troops for the immediate instigation of civil war.[1] On the day of the German invasion of Soviet Russia the Comintern instructed all communist parties in Europe to raise the standard of revolt, and ten days later the Yugoslav Communist Party was directed to organize guerilla units at once.[2] Political instructions were given in Moscow's short-wave broadcasts.[3]

It is obvious from the performance of the French and Italian communist partisans that they were instructed by Moscow to co-operate with the West. The reason is not far to seek.

'The war was still on, and Allied military commanders had to be obeyed. This was as important for Moscow as for Washington or London, and the leaders of the French (and Italian) communists were obliged to order their followers to give wholehearted support to the war effort.'[4]

The same considerations should have applied to Greece, since it was to be liberated by British troops, but this was not the case. On May 4, 1944, Sir Winston Churchill noted that 'evidently we are approaching a showdown with the Russians about their communist intrigues in Italy, Yugoslavia and Greece';[5] in his cable to President Roosevelt of June 23, 1944, he spoke of 'persuading the Russians to quit boosting EAM and ramming it forward with all their force',[6] and

---

[1] For the text see T. A. Taracouzio, *The Soviet Union and International Law*, New York 1935, pp. 439 and 441; and Dixon/Heilbrunn, *Communist Guerilla Warfare*, op. cit., pp. 24 and 25.

[2] Professor F. W. D. Deakin, *Great Britain and European Resistance*, loc. cit.

[3] Cf. L. de Jong, *The Allies and Dutch Resistance 1940–45*, Paper read at the Second International Conference etc., as above.

[4] Hugh Seton-Watson, *The Pattern of Communist Revolution, A Historical Analysis* London 1953, pp. 222 and 225.

[5] *The Second World War*, vol. v, p. 623; vol. vi, p. 64.

[6] *Ibid*, vol. vi, p. 68.

APPENDIX

he subsequently complained about the bad faith of the Russians who 'early in August... dispatched from Italy by a subterfuge a mission to ELAS in Northern Greece'.[1] Soviet promotion of civil war in Greece seems to have stopped only after Britain had paid 'the price we have to Russia for freedom of action in Greece';[2] this is a reference to the October 1944 agreement in Moscow on the respective British-Soviet predominance in the various Balkan countries, which gave to Britain 'ninety per cent. of the say in Greece'.[3] 'Stalin', Sir Winston remarked subsequently, 'adhered strictly and faithfully to our agreement of October.'[4]

While the Soviets, up to this point, promoted internecine struggle in Greece, they tried to stop it in Yugoslavia. In February 1942, Moscow admonished Tito for forming special Proletarian Brigades.

'Surely', the message went on, 'at the moment the basic, immediate task is to unite all anti-Nazi currents, smash the invaders and achieve national liberation.... It is difficult to agree that the London and the Yugoslav governments are siding with the invaders. There must be some great misunderstanding here. We honestly request you to give your tactics serious thought, and your actions as well, and make sure that on your side you have really done all you could to achieve a true united national front of all enemies of Hitler and Mussolini in Yugoslavia in order to attain the common aim—the expulsion of the invaders and would-be conquerors.'[5]

How does this policy square with the one adopted by the Soviets in Greece?

Sir Fitzroy Maclean has given an explanation.

'Tito was going ahead too fast for their taste. They had no wish to alarm their British and American allies unduly at this stage (October 1943); the time for that would come later. Nor was it for Tito to take the initiative in such matters. The Kremlin would decide in good time when and how a communist regime was to be established in Yugoslavia.'[6] This is certainly true, but one additional factor must

---

[1] *The Second World War*, vol. vi, p. 71.   [2] *Ibid*, vol. vi, p. 250.
[3] *Ibid*, vol. vi, p. 198.   [4] *Ibid*, vol. vi, p. 255.
[5] Quoted from Vladimir Dedijer, *Tito Speaks, op. cit.*, p. 178.
[6] *Disputed Barricade, op. cit.*, p. 251.

be borne in mind: Greece would not be liberated by the Red Army but Yugoslavia would, and a country liberated by the Red Army would be drawn into the Soviet orbit anyway. In fact, Stalin claimed in his post-war dispute with Tito that the latter did not gain power because of any special merits of the Yugoslav Communist Party but because 'the Soviet Army came to the aid of the Yugoslav people, crushed the German invaders, liberated Belgrade, and in this way created the conditions which were necessary for the Yugoslav Communist Party to achieve power'.[1]

Which prompted Tito to the rejoinder that Stalin held 'the belief that only occupation by the Red Army may bring the new socialist order'[2]—except in the case of Greece where the Soviet Army could not come to the rescue and, in Sir Winston's words, EAM had to be rammed forward with all force.

The ideological bond between Soviet Russia and the communist partisan movements in Europe was so strong that there was no need to ensure control over the partisans by other means. In fact, the Soviets sent a military mission to Tito only in February 1944, and to Greece even later, in August 1944. Soviet supplies did not reach the partisans in Yugoslavia until April 1944, Polish communist partisans were first supplied in the same month, Czechoslovakian partisans had to wait until July of that year and apparently nothing was sent to communist guerillas in Albania and Greece. Czechoslovakia also received, as we have mentioned in Chapter 8, officers with guerilla experience and about 400 partisans from Soviet Russia. For the rest Soviet support was moral.

Was the war effort hampered by the East-West differences in the guerilla theatres? It is obvious that the guerillas in Greece, Yugoslavia and Albania would have been more profitably employed had they only fought against the occupying powers and not also against each other.[3] This would also have prevented the Axis from trying to play off one partisan movement against the other, although it must be recognized that the relief obtained by the Germans and Italians in

---

[1] *The Soviet-Yugoslav Dispute*, London 1948, p. 51.
[2] *Marshal Tito Replies to French Journalist Louis Dalma*, Yugoslav Fortnightly, I, No. 17 (1950), quoted from Elliot R. Goodman, *The Soviet Design for a World State*, op. cit., p. 311.
[3] For particulars cf. O. Heilbrunn, *The Soviet Secret Services*, op. cit., pp. 101 f.

APPENDIX

this way was negligible; they always realized that those who had made accommodations with them would turn against them once more as soon as the situation permitted. More serious was the fact, mentioned by Colonel Woodhouse, that the Greek population became so disgusted with the constant strife that the German-supported government in Athens succeeded in recruiting many Greeks into their anti-partisan forces.[1] It is also one of the tragedies of the last war that the Soviet attitude made it impossible for the West to give more help to the Polish Home Army in its final struggle. Those who are inclined to measure resistance in terms of a profit and loss account may say that the communist contribution to partisan warfare in Europe was so great that the above entries on the debit side do not matter. There are others who reject so crude a reckoning.

There was another difference between East and West in the conception of partisan warfare. In the Eastern view partisan forces must be grouped in larger units and engage in combat action from start to finish to the extent shown in Soviet Russia and Yugoslavia. This was made clear in Stalin's famous order to the Soviet Russian People, issued early in July 1941, in which he asked for the formation of guerilla units 'to combat enemy troops' and to pursue and annihilate the enemy everywhere. Massive partisan strength and combat action is even more necessary where an imperialist war must be transformed into a civil war.

The British idea of partisan war was rather different. As Professor Deakin has pointed out, Britain thought of partisans operating in small groups and mainly engaged in intelligence and sabotage tasks, followed perhaps later by guerilla attacks on a larger scale when the front approached the partisan area.[2] The overriding consideration which determined partisan strategy was that the sacrifices which the population would have to bear must be commensurate with the contribution which guerilla activities would make to the war effort. The various countries were not therefore indiscriminately asked for a maximum partisan contribution. Its scope was decided individually for each country concerned, and its geographical location was the determining factor.

As a result partisan tasks in the various countries were more or less graduated on the following lines:

[1] In *The Greek Resistance 1942–44, op. cit.*, p. 385.
[2] In *Great Britain and European Resistance, loc. cit.*

1. Countries outside the future battle zones, such as Norway and Denmark. Since no Allied military assistance could be forthcoming, these countries were advised not to take overt action but to restrict their resistance activities to the prevention of the scorched earth policy and, as a secondary task, to undertake certain permitted sabotage missions.

2. Countries through which important enemy lines of communication ran, such as Greece and Yugoslavia. They were either given specific sabotage tasks (as were the Greek guerillas and those of Mihailovitch) or general sabotage directives (as were Tito's partisans).

3. Countries which could tie down enemy forces. Again Greece and Yugoslavia were the obvious choice. Before the Allied landing in Sicily, Greek guerillas had to give the impression that the landing would take place in Greece and carry out widespread attacks on the communication lines.[1] In Yugoslavia the same task was assigned to Mihailovitch's forces who, however, failed to carry it out.

4. Countries which would become battlefields, mainly France and Belgium. There was of course no question of drawing enemy forces into these countries. Large-scale guerilla activities were therefore postponed in these countries until immediately before the landing in Normandy and the approach of the front to Belgium.

All countries were expected to supply as much intelligence as possible and to carry out minor sabotage activities all the time.

These East-West differences in regard to the scope of partisan warfare did not arouse any misgivings, and the West especially had no reason for any. Communist partisans in western Europe carried out their assigned tasks, and the very considerably larger mission taken on by Tito was in line with the British conception of Yugoslavia's special contribution to the war effort.

Probably all students of partisan warfare are agreed that the partisan contribution to the outcome of World War II was not decisive and that the Allied armed forces would have won the war without the help of the Resistance. Yet there is no doubt either that on the whole the partisans gave valuable assistance to the Allied armies. Its true value can only be very approximately assessed, because partisan activities were so varied—killing enemy soldiers here, delaying others somewhere else, reporting enemy troop move-

[1] Cf. Colonel C. M. Woodhouse, in *The Greek Resistance, 1942–44*, loc. cit., p. 382.

APPENDIX

ments in one theatre and tying down a number of divisions in another, keeping up the morale of the population and accepting the surrender of a garrison. Furthermore many figures, especially those relating to partisan strength and enemy casualties, are controversial, regardless of the source from which they come. Two examples will suffice: 50,000 people are believed to have belonged to the Belgian Resistance in 1944, but today it is claimed that the number was 136,000;[1] and Colonel Woodhouse, an eye-witness of the final Greek guerilla operations against the Germans, describes the German casualty figures—5,000 wounded and captured—given in the British Official History of the war as 'absurdly inflated'.[2] There is the added difficulty that many figures have not been published at all. Finally, in the field of partisan intelligence, not all relevant facts have become known, especially as far as Soviet partisan intelligence is concerned.

The usefulness of research into the actual value of partisan warfare in World War II seems therefore to be limited. But in spite of all shortcomings it appears that a definite appreciation is possible if we limit the range of our inquiry to those countries where partisan operations and partisan intelligence were most active, that is, for operations to Yugoslavia and Soviet Russia and for intelligence to Soviet Russia and France.

As for operations in Yugoslavia and Soviet Russia, we have selected the year 1943 because, in spite of some uncertainties, the most reliable figures are available for this time. In Yugoslavia heavy fighting had been in progress through much of the year which saw the Fourth, Fifth and Sixth Enemy Offensives. The German forces were strong—we shall see in a moment how strong they were—and they were by any reckoning far superior to the partisan strength. Brigadier Maclean holds the considered opinion that the Germans, in spite of their superiority in matériel, could no longer hope to wipe out the partisans altogether in the course of a single offensive but that they could always dislodge the partisans from any position. The Germans therefore decided to deal with the partisans piecemeal.[3] The same ob-

[1] H. Michel, 'Rapport Général,' in *European Resistance Movements, 1939–1945*, *op. cit.*, p. 57, quoting the article by M. Lejeune in *Revue d' Histoire de la 2ème Guerre Mondiale* of July, 1958, as source, states that such an inflation is not peculiar to Belgium.
[2] *Loc. cit.*, p. 376.
[3] In *Disputed Barricade, op. cit.*, p. 248.

servation holds good in regard to the fighting in Soviet Russia; here too the Germans could clean up a sector if they had the troops available, but they were unable to eliminate the partisan menace in more than a limited area at any given time.[1] This nibbling process never showed any decisive results because the partisans in Yugoslavia and Soviet Russia could always more than make good their losses in men and arms.

It is therefore obvious that two factors indefinitely prolonged the partisan war in both countries: the one was the reservoir of partisan manpower and supplies and the other the near-balance of partisan and anti-partisan capability. We therefore propose to look now at the number of forces employed by each side in the two partisan theatres.

According to a German appreciation for the end of 1943, referred to be Sir Fitzroy Maclean,[2] the Germans had available in Yugoslavia:

14 German Divisions
2 German SS Regiments
5 non-German divisions under German command

Total 200,000 men
Bulgarians, Serbs and Croats   160,000 men

Grand total 360,000 men

According to the same appreciation the partisans had in fighting troops   110,000 men

This last figure is obviously too low because Sir Winston Churchill estimates the partisan strength at about this time at 200,000 and the Yugoslavs themselves put it officially at 250,000 for the autumn of that year.[3] The German figure can of course only be an estimate and one must therefore assume that the Yugoslav partisans tied down more than half as many enemy troops again.

We have also to rely on a German appreciation for Soviet Russia, because we know of no Soviet estimate of the relative strength of the opposing sides at any given time. The German Army Group

[1] Major Edgar M. Howell, *The Soviet Partisan Movement*, 1941-44, op. cit., p. 210.
[2] In *Disputed Barricade*, p. 248.
[3] Lieut.-Col. Brajuskovie-Dimitrye, *La Guerre de Libération Nationale en Yougoslavie* (1941-45), in European Resistance Movements 1939-45, op. cit., p. 336.

APPENDIX

North which fought in Soviet Russia has compiled a survey of the partisan situation in its forward area; it was a secret document, meant for internal use only, and while it too may be mistaken about the partisan strength it is most likely the best available source.[1]

On the German side there fought against the partisans in this theatre 16 German and 3 other battalions, 135 German and 22 other companies, and 87 other units, mostly German, of 'less than company' strength. Neither the individual nor the total strength of the anti-partisan forces is given, but if we assume that the 19 battalions had about 400 men each, the 157 companies about 120 men each, and the 87 other units about 50 men each, we arrive at a total strength of just over 30,000 men.

On the partisan side there fought 11 Lenin Brigades whose individual strength is given in the document. These forces total

14,300 men

There were also 7 cavalry brigades for each of which the strength is again given; they total   3,800 men

Finally, there were 10 other detachments, with a total strength of   4,950 men

23,050 men

The document also shows two more brigades without any indication of their strength, and 10 brigades are listed as dispersed or of unknown location. The total partisan strength in the forward area of Army Group North can approximately be estimated as between 25,000 and 30,000 men.

These partisans therefore tied down at least an equal number of Security Forces.

Approximately the same seems to hold good for Army Group Centre which also fought in Soviet Russia. According to post-war American estimates based on contemporary reports, the partisans in the Army Group area occupied the attention of nearly 100,000 German and allied troops.[2] As a Partisan Situation Map of this Army Group shows, 56,000 partisans were estimated to operate in this area on June 30, 1943. By September 30, 1943, the partisan forces had

[1] *Feldzug gegen die Sowjetunion der Heeresgruppe Nord, Kriegsjahr* 1943, *op. cit.*
[2] Stewart Alsop and Colonel S. B. Griffith: 'We Can Be Guerillas Too', *Saturday Evening Post*, December 2, 1950, pp. 32 f.

considerably increased; the Army Group estimated the strength of the known detachments at roughly 75,000 men—each detachment is assessed separately on the map. But there are also sixteen question marks on this map, indicating that the estimated strength of a particular detachment is doubtful or that no estimate for the strength of a detachment could be given. There is also one partisan detachment of 12,000 men which operated partly in this and partly in another area. By and large one can assume that the partisans in Army Group Centre area, as in the North, tied down security forces slightly more numerous than themselves.[1]

There is no doubt that most partisan actions inflicted damage upon the opposing forces. Some of the damage was severe, such as the dislocation of German communications by the French partisans at the time of the Normandy landing, the destruction of a vital railway viaduct by Greek partisans in October 1942, the liberation of the port of Antwerp by Belgian partisans in September 1944, and the surrender of the garrison of Genoa to Italian partisans in April 1945. Damage was also done by many thousands of individual actions, mining, ambushing, destroying, fighting, killing. But that by these actions the partisans, properly deployed in strategically important areas, could permanently tie down as many or more of an enemy who was often supported by tanks and aircraft, was probably the most impressive combat contribution of the partisans in World War II.[2]

Their second great contribution was in the field of intelligence. It is difficult to assess the value of Soviet partisan intelligence. It is evident that for strategic information the Soviets relied on their espionage networks in Japan and Switzerland, but it cannot be doubted that the partisans served well as field intelligence, especially after Army intelligence officers had been seconded to all partisan staffs in 1943. The scope was wide—the partisans were everywhere—their location ideal—behind the enemy's front—and their instructions were detailed—in the Field Service Regulations, the *Partisan Handbook*, the *Guide Book for Partisans* and so on.

'We can be almost certain that again and again Russian attacks were

[1] The two Partisan Situation Maps are reproduced in Major Edgar M. Howell's *The Soviet Partisan Movement 1941-44*, op. cit., opposite pp. 161 and 171.

[2] The last war has not shown whether partisans could survive for any length of time if they drew considerably more than their own numbers upon themselves, or where the breaking point is.

APPENDIX

mounted in those areas which partisan reports had indicated as vulnerable. The Russians during the war became expert in attacking the enemy's weakest points: the small front-line gaps in the winter of 1941-2, the front held by German satellite troops at the beginning of the Stalingrad battle; and if there was neither gap nor satellite, it was almost always the seam between two enemy formations which the Red Army selected for its breakthrough attempts.... There was only one source which could consistently direct the Red Army against the weakest link of the enemy front line, and this task ... was entrusted to the partisans behind the line.'[1]

We are of course better informed about the value of French partisan intelligence.

'In fact, the day the battle (in France) began', says General de Gaulle, 'all the German troop emplacements, bases, depots, landing fields and command posts were precisely known, the striking force and equipment counted, the defence works photographed, the minefields spotted.... Thanks to all the information furnished by the French resistance, the Allies were in a position to see into the enemy's hand and strike with telling effect.'[2]

These words speak for themselves; no finer testimonial could be given.

'The second world war aggravated the crisis of the colonial system as expressed in the rise of a powerful movement for national liberation in the colonies and dependencies.... The attempt to crush the national liberation movement by military force now increasingly meets armed resistance on the part of colonial peoples and produces prolonged colonial wars.'[3]

In these words Zhdanov addressed the inaugural Cominform Conference in 1947. The Indo-Chinese, Philippino and Malayan Communist movements which embarked on wars of liberation in the post-

---

[1] Dixon/Heilbrunn, *Communist Guerilla Warfare*, op. cit., p. 188.
[2] General de Gaulle, *War Memoirs*, vol. ii, *Unity, 1942-44*, London 1959, p. 282.
[3] Quotation from Elliot R. Goodman, *The Soviet Design for a World State*, op. cit., p. 313.

war years, had all fought during World War II against the Japanese. So had Mao Tse-tung; and his guerillas, almost a spent force before the anti-Japanese war, had gathered great strength during this fight, enough at any rate to make them confident of winning the resumed civil war. The North Koreans had Mao's armed support. The Greek Communists who had fought against the Germans during the war were the nucleus of the forces which with Yugoslavia's support started the civil war in 1947, and Colonel Grivas has claimed to have acquired his guerilla experience by fighting the Germans in Greece. Guerilla war, a minor adjunct in 1914–18, and then, in 1939–45, a second front behind each front, had now become the vehicle of national-revolutionary action. Of all the more troublesome post-war movements for 'national liberation' only those in Africa—Kenya, Tunisia and Algeria—could not claim to have found their strength in World War II exploits against the invaders.

None of these post-war movements can invoke the name of the Resistance.

They are not the consequences of the Resistance but of World War II itself.

BIBLIOGRAPHY

ALASTOS, DOROS: *Cyprus Guerilla, Grivas. Makarios and the British.* London, 1960.
ALLEN, W. E. D.: 'Gideon Force.' In Irwin R. Blacker (ed.): *Irregulars, Partisans, Guerillas.* New York, 1954.
ALSOP, STEWART, and Colonel S. B. GRIFFITH: 'We Can Be Guerillas Too.' *Saturday Evening Post*, December 2, 1950.
AMERY, JULIAN: *Sons of the Eagle. A Study in Guerilla War.* London, 1948.
ANISIMOV, OLEG: 'The German Occupation in Northern Russia during World War II. Political and Administrative Aspects.' Research Program on the U.S.S.R. New York City, 1945.
BALCOS, ANASTASE: 'Guerilla Warfare.' In *Military Review*, March 1958. Reprinted in *Military Digest*, No. 41, April 1959.
BAND, CLAIRE and WILLIAM: *Dragon Fangs. Two Years with Chinese Guerillas.* London, 1948.
BARKER, DUDLEY: *Grivas, Portrait of a Terrorist.* London, 1959.
BEEBE, Lt-Colonel JOHN E.: 'Beating the Guerillas.' In *Military Review*, vol. xxxv, December 1955.
BEGIN, MENACHEM: *The Revolt.* London, 1951.
BERRY, Brigadier-General: 'Statement by U.K. Representatives.' In *European Resistance Movements 1939–45*, Oxford etc., 1960.
BRAJUSKOVIE-DIMITRYE, Lt.-Colonel: 'La Guerre de Libération en Yougoslavie (1941–5).' In *European Resistance Movements 1939–45*, Oxford etc., 1960.
BRAZIER-CREAGH, Brigadier K. R.: Malaya. Lecture reprinted in *Royal United Service Institution Journal* 1954, p. 175.
BYFORD-JONES, W.: *Grivas and the Story of Eoka.* London, 1959.
CASTEX, Amiral: 'Les enseignements de la Guerre d'Indochine.' In *Revue de Défense Nationale* 1955, vol. xxi.
CAUCHETIER, Lieutenant R.: Na-Sam. In *Forces Aériennes Françaises*, Indochine 1953.

CHAPELLE, DICKEY: 'How Castro Won.' In *Marine Corps Gazette*, February 1960.
CHAPMAN, Colonel F. SPENCER: *The Jungle is Neutral*. London, 1949.
CHASSIN, Général L.-M.: *L'Ascension de Mao Tse-tung* (1921–45). Paris, 1953.
—*La Conquête de la Chine par Mao Tse-tung* (1945–9). Paris, 1952.
—'Guerre en Indochine.' In *Revue de Défense Nationale*, vol. xvii, 1953.
CHINA WHITE PAPER. A Summary with Commentary of the Department of State's 'U.S. Relations with China' by Francis Valeo. Washington, 1949.
CHURCHILL, Sir WINSTON: *The Second World War*. Vol. iii, London, 1950; vol. iv, London, 1951; vol. v, London, 1952; vol. vi, London, 1954.
CLUTTERBUCK, Lt.-Colonel R. L.: 'Bertrand Stewart Essay 1960.' In *The Army Quarterly*, January 1961, pp. 164 f.
CORFIELD, F. D.: *Historical Survey of the Origins and Growth of Mau Mau*. Colonial Office, H.M. Stationery Office, London, 1960.
CROIZAT, Colonel VICTOR J.: 'The Algerian War.' First published in *Marine Corps Gazette* and reprinted in *An Cosantóir*, 1958, p. 18.
CROZIER, BRIAN: *The Rebels*. London, 1960.
DACH BERN, HAUPTMANN H. VON: Der totale Widerstand, Kleinkriegsanleitung für jedermann. *Schriftenreihe des Schweizerischen Unteroffiziersverbandes*, Biel, 1958.
DALLIN, D. J.: *Soviet Espionage*. New Haven, 1955.
DEAKIN, Professor F. W. D.: Great Britain and European Resistance. Paper read at the Second International Conference on the History of the Resistance, Milan, March 26/29, 1961.
DEDIJER, VLADIMIR: *Tito Speaks*. London, 1953.
DEMANGE, Colonel: 'La Guérilla.' In *Revue Militaire Générale*, February 1960.
DIXON, Brigadier C. A., and O. HEILBRUNN: *Communist Guerilla Warfare*. London, Allen & Unwin, 1954; New York, Praeger 1955; Paris, 1956; Frankfurt (Main)-Berlin, 1956.
DOUGHERTY, JAMES E.: 'The Guerilla War in Malaya.' In *U.S. Naval Institute Proceedings*, vol. 84, No. 9, September 1958.
L'ECOLE D'APPLICATION DE L'INFANTERIE DE SAINT-MAIXENT:

BIBLIOGRAPHY

'Manual on Anti-Guerilla Warfare.' In *Bulletin Militaire*, December 1957.

ERSKINE, General Sir GEORGE: 'Kenya—Mau Mau.' Lecture reprinted in *Royal United Service Institution Journal* 1956, pp. 11 f.

ESSON, Major D. M. R.: 'The Secret Weapon—Terrorism.' In *The Army Quarterly*, 1959, p. 179.

FALL, Professor BERNARD: 'Das Ende der Kampfgruppe 100.' In *Wehrwissenschaftliche Rundschau* 1960, p. 611.

—*Street Without Joy. Indochina at War, 1946–54*. Harrisburg 1961.

FARRAN, JEAN: 'La Leçon de Dien Bien Phu.' In *Bulletin Militaire*, August 1956.

FARRAN, ROY: *Operation Tombola*. London, 1960.

FEDERATION OF MALAYA: *Annual Reports*, 1954, 1955, and 1956.

FEDOROV, A.: *L'Obkom clandestin au travail*. Paris, 1951.

FELDZUG GEGEN DIE SOWJETUNION DER HEERESGRUPPE NORD, *Kriegsjahr 1943. Oberkommando der Heeresgruppe Nord*, December 24, 1944.

FLICKE, W. L.: *Agenten funken nach Moskau*. Kreuzlingen, 1954.

FOXLEY-NORRIS, Wing Commander C. N.: 'The Use of Airpower in Security Operations.' In *Royal United Service Institution Journal* 1954, p. 555.

FULLER, Colonel FRANCIS F.:'Mao Tse-tung, Military Thinker.' In *Military Affairs*, 1958, p. 145.

GARTHOFF, RAYMOND L.: *Soviet Military Doctrine*. Glencoc, Illinois, 1953.

DE GAULLE, General: *War Memoirs*. Vol. ii, Unity, 1942-4. London, 1959.

GERMAN COUNTER-INTELLIGENCE ACTIVITIES IN OCCUPIED RUSSIA (1941–4). Office of the Chief of Military History, U.S. Department of the Army, Washington, n.d.

GOODMAN, ELLIOT R.: *The Soviet Design for a World State*. New York, 1960.

GOURLAY, Major B. I. S.: 'Terror in Cyprus.' In *Marine Corps Gazette*, September 1959.

GREVECOEUR, Colonel DE: 'La guerre du Vietminh.' In *Tropique*, June 1953.

GUERILLA SELON L'ECOLE COMMUNISTE.Viet-minh Directives published by the Etat-Major de la Force Publique in Léopoldville in *Bulletin Militaire*, June and August, 1955.

GUEVARA, Major E., *La Guerra de las Guerrillas*, Havana, 1960.
HANRAHAN, GENE Z.: *The Communist Struggle in Malaya*. New York, 1954.
—'The Chinese Red Army and Guerilla Warfare.' In *U.S. Army Combat Forces Journal*, February 1951.
—'Guerilla Warfare.' In *Marine Corps Gazette* 1956, p. 31.
HARRISON, D. I.: *These Men are Dangerous*. London, 1957.
HARRISON, TOM: *World Within. A Borneo Story*. London, 1959.
HAWEMANN, WALTER: *Achtung! Partisanen*. Hanover, 1953.
HEILBRUNN, OTTO: *Communist Guerilla Warfare*, see DIXON, Brigadier C. A., and O. HEILBRUNN.
—*The Soviet Secret Services*. London, Allen & Unwin, 1956; New York, Praeger 1956; Frankfurt (Main)-Berlin, 1956.
—*Partisanenbuch*. Zürich, 1960.
HENDERSON, IAN: *The Hunt for Kimathi*. London, 1958.
HOGARD, Commandant J.: 'Guerre révolutionnaire ou Révolution dans l'art de la guerre.' *Revue de Défense Nationale*, Dec. 1956.
—'Le soldat dans la guerre révolutionnaire.' In *Revue de Défense Nationale*, February 1957.
—'Guerre révolutionnaire et pacification.' In *Revue Militaire d'Information*, January 1957.
HOWELL, Major EDGAR M.: *The Soviet Partisan Movement 1941–4*. Department of the Army Pamphlet, Washington, 1956.
HULL, CORDELL: *The Memoirs of*, vol. ii. London, 1948.
IGNATOW, P. K.: *Partisanen*. Berlin, 1953.
DE JONG, L.: 'The Allies and Dutch Resistance 1940–5.' Paper read at the Second International Conference on the History of the Resistance. Milan, March 26/29, 1961.
KESSELRING: *The Memoirs of Field-Marshal*. London, 1953.
KITSON: Major FRANK: *Gangs and Counter-Gangs*. London, 1960.
VON KNIERIEM, August: *Nürnberg. Rechtliche und menschliche Probleme*. Stuttgart, 1953.
KOGAN, NORMAN: 'American Policies towards European Resistance Movements.' Paper read at the Second International Conference on the History of the Resistance. Milan, March 26/29, 1961.
KOVIC-DIMITRYE, Lt.-Colonel BRAJUS: 'La Guerre de libération Nationale en Yvugoslavie (1941–5).' In *European Resistance Movements 1939–45*. Oxford etc., 1960.

BIBLIOGRAPHY

KOVPAK, Major-General S. A.: *Our Partisan Course*. London, 1947.
KRAJINA, Professor VLADIMIR: 'La Résistance tchécoslovaque.' In *Cahiers d'Histoire de la Guerre*, February 1951, No. 3.
KRIEGSHEIM, Major HERBERT: *Getarnt, Getäuscht und doch Getreu. Die geheimnisvollen 'Brandenburger'*. Berlin, 1958.
KUTGER, Lt.-Colonel JOSEPH P.: 'Irregular Warfare in Transition.' In *Military Affairs*, vol. xxiv, No. 3, 1960.
LAUZIN, Général CH.: 'Opérations en Indochine.' In *Forces Aériennes Françaises*, March 1955.
LEIGH, IONE: *In the Shadow of Mau Mau*. London, 1954.
LEVERKUEHN, PAUL: *Der geheime Nachrichtendienst der deutschen Wehrmacht im Kriege*. Frankfurt am Main, 1957.
MACLEAN, Brigadier Sir FITZROY: *Disputed Barricade*. London, 1957.
MCQUILLEN, Colonel J. F.: 'Indochina.' In *Marine Corps Gazette* 1955.
VON MANSTEIN, Field-Marshal: Transcript in the Court-Martial case.
MANUAL FOR WARFARE AGAINST BANDS: see: Richtlinien etc.
MAO TSE-TUNG: 'On the Protracted War.' In *Selected Works of Mao Tse-tung*, vol. ii, London, 1954.
—'Strategic Problems of Guerilla War.' *Ibidem*.
—Interview with the British Correspondent James Bertram. *Ibidem*.
—*Aspects of China's Anti-Jap Struggle*. Bombay, 1948.
—'Problems of War and Strategy.' In *Selected Works*, vol. ii.
MARLIERE, Major BEM L.: 'Quelques leçons á tirer des Campagnes du Laos.' In *Bulletin Militaire* 1955, p. 559.
MERYE, JEAN: 'Considérations militaires sur la Guerre d'Algérie.' *Revue de Défense Nationale*, May 1959.
MICHEL, HENRI: 'Rapport Général.' In *European Resistance Movements, 1939-45*. Oxford etc., 1960.
MIERS, Lt.-Colonel RICHARD: *Shoot to Kill*. London, 1959.
MIHAILOVIC, *The Trial of*. Stenographic Records and Documents. Belgrade, 1946.
MIKSCHE, Lt.-Colonel F. O.: *Secret Forces. The Technique of Underground Movements*. London, 1950.
MILLER, H.: *Menace in Malaya*. London, 1954.
MITAUX-MAROUARD, Commandant: 'La défense des bases

aériennes en Indochine.' In *Forces Aériennes Françaises*, Indochine 1953.

MONTGOMERY, Field-Marshal Viscount: *Memoirs*. London, 1958.

MURRAY, Colonel J. C.: 'The Anti-Bandit War.' In *Marine Corps Gazette*, January to May, 1954.

MYERS, Brigadier E. C. W.: *Greek Entanglement*. London, 1955.

NOLL: 'The Emergency in Malaya.' In *The Army Quarterly*, April 1954.

THE NORWEGIAN HOME GUARD. A Survey of. Issued by the Inspector-General of the Norwegian Home Guard. Oslo, May 1955.

NUREMBERG TRIALS. Hostage Case, Case No. VII in the Subsequent Trials.

O'BALLANCE, Major EDGAR: 'The Algerian Struggle.' In *The Army Quarterly*, October 1960.

PARET, PETER: 'The French Army and la guerre révolutionnaire.' In *Royal United Service Institution Journal*, vol. 104, 1959, p. 60.

REDELIS, VALDIS: *Partisanenkrieg. Die Wehrmacht im Kampf*. Heidelberg, 1958.

RICHTLINIEN DES OBERKOMMANDOS DER WEHRMACHT FUER DIE BANDENBEKAEMPFUNG vom 6. Mai 1944.

RIGG, Colonel ROBERT B.: *Red China's Fighting Hordes*. Harrisburg, 1952.

—'Red Parallel, The Tactics of Ho and Mao.' In *Army Combat Forces Journal*, January 1955.

RILEY, Major DAVID: 'French Helicopter Operations in Algeria.' In *Marine Corps Gazette*, February 1958.

ROBINSON, Major R. E. R.: 'Reflections of a Company Commander.' In *The Army Quarterly*, vol. LXI, No. 1, October 1950.

SAUNDERS, HILARY ST GEORGE: *Royal Air Force*, 1939-45. Vol. iii, *The Fight is Won*. London, 1954.

SEETH, Flight-Lieutenant D. R.: 'The Employment of Air Power in Malaya.' In *Indian Air Force Quarterly*, October 1954.

SETH, RONALD: *The Undaunted. The Story of Resistance in Western Europe*. London, 1956.

SETON-WATSON, HUGH: *The Pattern of Communist Revolution. A Historical Analysis*. London, 1953.

SLATER, Group Captain K. R. C.: 'Air Operations in Malaya.' In *Royal United Service Institution Journal* 1957, p. 380.

BIBLIOGRAPHY

SOUYRIS, Capitaine ANDRÉ: 'Un procédé efficace de Contre-guérilla.' In *Revue de Défense Nationale*, 1956, p. 686.

—'Les conditions de la parade et de la riposte à la Guerre révolutionnaire.' In *Revue Militaire d'Information*, February/March, 1957.

SOVIET-YUGOSLAV DISPUTE, THE. London, 1948.

TARACOUZIO, T. A.: *The Soviet Union and International Law*. New York, 1935.

TAYLOR, JOE G.: 'Air Support for Guerillas on Cebu.' In *Military Affairs*, vol. xxiii, No. 3, 1959.

TEMMERMAN, JEAN: 'La Résistance Belge.' In *L'Armée, La Nation*, January 1955.

TESKE, H.: 'Partisanen gegen die Eisenbahn.' In *Wehrwissenschaftliche Rundschau*, Oktober 1953, Heft 10.

THURSBY, Major P. D. F.: 'Helicopter Operations in Cyprus.' In *The Suffolk Regimental Gazette*, Summer 1957.

VACCARINO, G.: 'Le Mouvement de Libération Nationale en Italie (1943–5).' In *Cahiers d'Histoire de la Guerre*, February 1951, No. 3.

VILLA-REAL, Lt.-Colonel LUIS A.: 'Huk Hunting.' In *Army Combat Forces Journal*, November 1954.

VOLLGRAFF, P. D.: 'La Résistance en Hollande.' In *Cahiers d'Histoire de la Guerre*, February 1951, No. 3.

WARNER, DENIS: *Out of the Gun*. London, 1956.

WILLOUGHBY, Major-General C. A.: *Sorge: Soviet Master Spy*. London, 1952.

WOODHOUSE, Colonel C. M.: *Apple of Discord*. London, 1948.

—'The Greek Resistance 1942–4.' In *European Resistance Movements 1939–45*. Oxford etc., 1960.

—Book review in: *The Twentieth Century*, June 1954.

XIMENES: 'La guerre révolutionnaire et ses données fondamentales.' In *Revue Militaire d'Information*, February/March 1957.

ZELENIKA, Colonel ILIJA: 'The Yugoslav Airforce.' In *Air Power*, Spring, 1954.

INDEX

A

Air Force, 123 f.
Albania, 15, 24, 30, 46, 102, 123
Algeria, 23, 128, 134, 135, 136, 137, 152, 156, 161, 165
Allen, W. E. D., 113
Alsop, S., 183
Ambush, 67, 92, 107, 108, 133
Amery, Julian, 23
Anti-partisans:
  Air Force, co-operation with, 125 f., 128 f.
  Ambush, 67, 104, 133
  Armoured Trains, 104 f.
  Band Situation Maps, 102
  Base Areas: Own, 99—Partisans', 45, 51, 66
  Basic Rules, 169
  Battue Shooting, 109
  Counter-bands, 69 f., 148
  Egg-beater, 104
  Encirclement, 65, 71 f., 126, 133
  Ferret Forces, 67, 103
  Home Guard, 51, 153, 157, 166
  Intelligence, 32 f., 69 f., 131
  Isolation of Partisans, 67 f.
  Jagdkommandos, 65, 68 f., 100
  Organization, 101
  Partridge Drive, 109
  Patrol Sweep, 104
  Pillbox Psychology, 36, 58, 64
  Propaganda among Population, 36, 155
  Protection of the Population, 35, 66, 153
  Psychological Warfare, 37, 135, 149, 156
  Short War, 166
  Support of the Population, 34, 160, 163, 166
  Surprise, 133, 136
  Surprise Attack, 103
  Tactics, 99 f.
  Tanks, 105, 142
  Techniques, 108 f.
  Treatment of Partisans, 143 f.
  Treatment of the Population, 36, 149 f.

Wars in:
  Albania, 102
  Algeria, 128, 134, 135, 136, 152, 156, 161, 165, 167
  China, 55 f., 64, 66
  Cyprus, 36, 102, 128, 132, 133, 135, 136, 161, 166, 167
  Greece, 102, 128, 129 f.
  Indo-China, 33, 36, 60 f., 70, 99, 100, 102, 128, 129, 130, 132, 134, 135, 150, 154, 163
  Kenya, 69 f., 132, 135, 137, 153, 162, 163, 167
  Malaya, 36, 67, 128, 130, 132, 133, 135, 136, 137, 153, 155, 157, 163
  Philippines, 155
  Soviet Russia, 65, 102, 126
  Yugoslavia, 71, 101, 105, 126
Armoured Trains, 104 f.

B

Barker, Dudley, 23, 37, 76, 132
Base Areas, 41 f., 45, 51, 53, 66, 80, 89, 91, 99, 159, 161
Battue Shooting, 109
Beebe, Lt-Col J. E., 73, 113
Begin, Menachem, 25
Belgium, 30, 112, 114, 180
Berry, Brig-General, 125
Black River, 62, 63
Blacker, I. R., 113
Bor, General T., 112
Borodin, 16, 41, 60
Brandenburg Division, 30, 120, 121
Brajuskovic-Dimitrye, Lt-Col, 182
Brazier-Creagh, Brig K. R., 157
Briggs Plan, 43, 67
Bryansk, 65

C

Castex, Amiral, 64
Castro, Fidel, 42, 44, 146, 160
Cauchetier, Lt R., 134
Chang Chun, 57, 66
Chapelle, Dickey, 146

Chassin, Général L-M., 35, 40, 55, 58, 59
Chenchow, 59
Chengteh, 59
Chetniks, 19, 30, 73
Chiang Kai-shek, 40, 53, 57, 59
China, 15, 16, 23, 35, 40 f., 43, 45, 48 f., 53 f., 62, 63, 64, 66, 67, 113, 117, 119, 160
China White Paper, 57, 59
Chindits, 120
Chunking, 55
Churchill, Sir Winston, 17, 174, 175, 176, 177
Clutterbuck, Lt-Col R. L., 104
Cochin-China, 62
Corfield, F. D., 23, 32
Counter-bands, 69, 148
Croizat, Col V. J., 138
Crozier, Brian, 145, 152, 154
Cuba, 42, 44, 160
Cyprus, 23, 34, 35, 36, 76, 102, 128, 132 f., 135, 136, 161, 166, 167
Czechoslovakia, 30, 125, 178

D

Dach Bern, Hauptmann H. von, 107, 139
Dallin, D. J., 17
Darling, General, 136
Deakin, Prof F. W. D., 174, 176, 179
Dedijer, V., 71, 73, 126, 177
Demange, Colonel, 114, 140
Denning, General Sir R. F. S. 117
Dien Bien Phu 63, 64, 66, 93, 100, 130, 134
Dixon, Brigadier C. A., 18, 147, 176, 185
Dougherty, James E., 67

E

L'Ecole d'Application de l'Infanterie de Saint-Maixent, 43, 101
Economic Warfare, 37 f.
Egg-beater, 104
ELAS, 24, 26, 27, 28, 174, 177
Encirclement:
   by anti-partisans, 65, 71 f., 108 f., 126, 133
   by partisans, 89 f.
Erskine, General Sir George, 102, 135, 137, 153
Esson, Major D. M. R., 151, 153

F

Fall, Prof B., 62, 64, 70, 100
Farran, Jean, 155
Farran, Roy, 140
Fedorov, A., 117
Ferret Forces, 67, 103
Field Service Regulations of the Red Army, 75, 78, 98, 138
Flicke, W. L., 17
Foxley-Norris, Wing Commander C. N., 129, 130, 132
France, 15, 30, 112, 114, 121, 180
Fuller, Colonel F. F., 147

G

Galen, General, 47
Garthoff, R., 125
Gaulle, Général de, 173, 185
German Air Force, 125 f.
Germany Army, xiii, 30, 65, 68, 69, 70, 102, 108 f., 113, 120, 124, 126, 147, 150, 151, 181
Gia-Hoi, 63
Giap, General, 43, 64
Goodman, Eliot R., 40, 178, 185
Gourlay, Major B. I. S., 152, 166
Greece, 15, 23, 24, 26, 28, 30, 34, 35, 46, 49, 51, 66, 102, 129, 143, 163, 165, 173, 174, 176, 178, 179, 180
Grevecoeur, Colonel de, 118
Griffith, Colonel S. B., 183
Guerillas, *see* Partisans
Guerre révolutionnaire, 159 f.
Guevara, Major E., 42

H

Hague Convention on Land Warfare, 143
Hankow, 55
Hanoi, 60, 63, 64
Hanrahan, Gene Z., 31, 38, 113, 118, 140
Harrison, D. I., 120
Hawemann, W., 65
Heilbrunn, Otto, 18, 22, 39, 77, 98, 147, 172, 176, 178, 185
Helicopter, 128, 131, 133, 134, 135, 136, 137, 166, 167
Henderson, Ian, 70
Himmler, H., 33, 102
Ho Chi-minh, 60, 62, 146, 160
Hoa Binh, 62
Hogard, Commandant J., 43, 157, 159, 162

INDEX

Holland, 22, 30
Home Guard, 51, 111, 153, 157, 166
Hongkong, 58
Howell, Major E. M., 18, 19, 65, 113, 124, 151, 152, 182, 184
Hsuchow, 59
Hull, Cordell, 17

I

Ignatow, P. K., 117
Indo-China, 23, 26, 33, 34, 36, 43, 47, 53, 60 f., 66, 70, 78 f., 99, 100, 102, 113, 118, 129, 130, 132, 134, 135, 138, 145, 150, 154, 160
Indonesia, 160
Initiative, see Partisans
Intelligence, see Anti-Partisans and Partisans, Intelligence
Irgun, 25
Italy, 22, 25, 27, 30, 114, 120, 121, 125

J

Jagdkommandos, 65, 68 f., 100, 167
Japan, 40, 47, 55, 60, 123, 147
Jebel, Akhdar, 134
Joffe, Adolf, 16
de Jong, L., 176

K

Kaifeng, 58
Kaobang, 62
Kenya, 23, 27, 69 f., 129, 132, 135, 137, 160, 163, 167
Kesselring, F. M., 114
Kimathi, 70, 129
Kirin, 57
Kitson, Major Frank, 69
Knieriem, A. von, 144
Kogan, N., 173
Korea, 73, 112, 119, 129, 170
Kovic-Dimitrye, Lt-Col B., 20
Kovpak, Major-General S. A., 117, 136
Krajina, Prof V., 17
Kriegsheim, Major Herbert, 30, 121
Kutger, Lt-Col J. P., 67

L

Laichau, 63
Langson, 62
Lao Kai, 62
Laos, 63, 64
Lauzin, Général Ch., 139
Leigh, Ione, 129

Leverkuehn, Paul, 30
Long Range Desert Group, 120, 139
Luftwaffe, see German Air Force

M

MacArthur, General D., 55
Maclean, Brigadier Sir Fitzroy, 35, 71, 73, 126, 175, 177, 181, 182
McQuillen, Colonel J. F., 38, 132
Malaya, 21, 23, 24, 26, 29, 31, 34, 35, 36, 37, 38, 42, 47, 60, 67, 73, 128, 130 f., 135, 136, 137, 153, 155, 157, 163
Manchuria, 48, 55, 56, 57, 59
Manstein, F.-M. E. von, 144
Manual for Warfare against Bands (German), 33, 36, 65, 69, 71, 76, 102 f., 104, 109, 128, 147, 150, 170
Mao Tse-tung, 16, 17, 35, 36, 40, 42, 44, 45, 46, 47, 56, 58, 59, 67, 80, 145, 146, 160, 164, 165, 168, 169
Marlière, Major L., 100
Mau Mau, see Partisans, Kenya
Michel, H., 181
Mihailovitch, General, 15, 19, 20, 21, 27, 30, 112, 114, 144, 151, 172, 173
Miksche, Lt-Colonel, F. O., 171
Militia, see Home Guard
Miller, H., 24, 133
Min Yuen, 31
Minsk, 65
Mitaux-Marouard, Commandant, 139
Montgomery, Field-Marshal Viscount, 151
Mukden, 57, 59
Murray, Colonel J. C., 129
Myers, Brigadier E. C. W., 115

N

Na-Sam, 63, 130, 134
Nanking, 55, 59
Nghia-Lo, 63
Nin Binh, 43
NKVD, 18, 26, 117, 125
Noll, 102
Norwegian Home Guard, 111, 116
Nuclear War, 32, 140 f.

O

O'Ballance, Major E., 152
Office of Strategic Services, 28, 116, 124
Organizational Charts, 26 f.
Orsha, 65

P

Palestine, 23, 25, 159
Paret, Peter, 160, 163

PARTISANS:
- Advance Preparation, 15, 115 f.
- Aggressiveness, 82
- Ambush, 92, 107
- Attack on Convoys and Railways, 107, 108
- Auxiliary Partisans, 39
- Bases, 41 f., 45, 51, 53, 66, 80, 89, 91, 96, 159, 161
- Co-operation with Air Force, 123 f.
- Co-operation with Army, 18, 23, 75, 112 f.
- Economic Warfare, 37 f.
- Encirclement, 89 f.
- Fifth Column, 95, 138
- Harassment, 88
- Inception:
  - Aims, 21 f.
  - Recruiting, 25 f.
  - Rival Movements, 29 f.
  - Structure, 23, 26 f.
  - Supplies, 28
- Independent Partisans, 39, 52
- Initiative, 56, 80 f., 83, 85, 86, 88, 96, 97
- Intelligence, 18, 19, 75 f., 87, 90, 113, 119, 138, 141, 184 f.
- Legal Status, 143 f.
- Operational Aspects, 53 f.
- Organization, 21 f., 118 f.
- Organizational Charts, 26 f.
- Partisan Handbook (Soviet), 78, 98, 108
- Partisan Manual (Indo-Chinese), 78 f.
- Population, 34, 46, 82, 86 f., 89, 90 f., 98, 145 f., 159
- Propaganda, 36, 44, 90, 95, 145, 159
- Protracted War, 44, 164 f.
- Regular Forces, Raising of, 47 f. 60
- Regular Forces, Relations with, 111 f.
- Revolutionary Movements, 16, 24, 32, 40 f., 78 f., 99, 159 f.
- Sabotage, 89, 98, 138
- Secrecy, 84, 87
- Speed, 84
- Strategy, 39, 40 f., 168
- Support of the Population, 34, 46, 87, 90, 145
- Surprise Attack, 94
- Tactics, 39, 67, 78 f., 168
- Techniques, 107 f.

PARTISANS (cont.)
- Terror, 145
- Treatment of Enemy, 146
- Wars in:
  - Albania, 15, 23, 30, 123
  - Algeria, 23, 165
  - Belgium, 30, 112, 114, 180
  - Borneo, 112
  - Burma, 112, 125
  - China, 15, 16, 35, 40 f., 43, 45, 48 f., 53 f., 64, 66, 67, 73, 113, 117, 119, 121
  - Cuba, 44
  - Cyprus, 23, 35, 36, 76
  - Czechoslovakia, 30, 125, 178
  - Ethiopia, 113
  - France, 15, 30, 112, 114, 172, 176, 180, 185
  - Greece:
    - 1941–44: 15, 24, 27, 28, 30, 33, 114, 143, 173, 174, 178, 179
    - Post-war: 23, 24, 28, 34, 46, 49, 51, 52, 66, 165
  - Holland, 22, 36, 112
  - Indo-China, 23, 26, 34, 36, 43, 47, 53, 60 f., 66, 78 f., 99, 113, 118, 121, 125, 134, 138, 145, 150
  - Italy, 21, 22, 25, 27, 30, 114, 176
  - Kenya, 23, 27, 33, 34, 35, 69 f., 129
  - Korea, 112, 119
  - Malaya:
    - 1941–45: 21, 26, 33, 112, 125
    - Post-war: 24, 29, 31, 34, 36, 37, 42, 44, 47, 73
  - Palestine, 23, 25
  - Philippines, 15, 30, 112, 123 f.
  - Poland, 30, 112, 125, 178, 179
  - Soviet-Russia, 15, 16, 17 f., 23, 25, 28, 66, 73, 75, 113, 118, 119, 121, 124, 144, 182 f., 184 f.
  - Tunisia, 23
  - Yugoslavia:
    - Mihailovitch, 15, 19, 22, 23, 27, 30, 112, 114, 144, 172, 174, 180
    - Tito, 15, 16, 19, 20, 26, 30, 45, 47, 48 f., 66, 71, 121, 123, 125, 126, 143, 144, 161, 172, 174, 177, 180, 181 f.
- World War II:
  - Allied Assistance, 21, 28, 112, 123 f., 172 f., 178
  - Allied Differences, 173, 176, 178 f.

INDEX

PARTISANS (*cont.*)
    Contribution to War Effort, 180 f.
    Functions, 179 f.
    Overall Control, 172 f., 178
    Reasons for widespread Activities, 171 f.

Patrol Sweeps, 104
Peking, 53, 55, 57, 59
Peng Teh-Huai, 58
Philippines, 15, 30
Poland, 30, 112, 125, 178, 179
Ponomarenko, 18
Population:
    Protection of, 35, 36, 66, 153
    Support of, 34 f., 155 f., 162, 166
    Treatment of, 36, 149 f.
Protracted War, 44, 164
Psychological Warfare, 37, 135, 149, 156

R

RAF, 68, 161 f.
Red Air Force, 124, 125
Red Army, 18, 19, 113, 116, 121
Red River Delta, 60 f., 154
Redelis, Valdis, 140
Richtlininen für die Bandenbekämpfung, *see* Manual for Warfare against Bands
Rigg, Colonel R. B., 53, 66, 100
Riley, Major D., 133, 136, 138, 168
Robinson, Major R. E. R. 73

S

SAS, 113, 120, 133, 134, 139
Saunders, Hilary St George, 123
Seeth, Lieutenant D. R., 131, 135
Seth, Ronald, 22, 25
Seton-Watson, Prof Hugh, 176
Shanghai, 53
Sian, 59
Singapore Special Training School, 21, 26
Slater, Group Captain K. R. C., 131, 135, 136, 137
Souyris, Capitaine A., 146, 154, 161, 163
Soviet Partisans, xiii, 15, 16, 17, 23, 25, 28, 66, 73, 74, 113, 118, 119, 124, 144, 183 f.
Special Boat Squadron, 113
Special Forces, 120 f.
Special Operations Executive, 21, 28, 116, 124, 174
Stalin, 18, 177, 179
Surprise Attack and Hunt, 103 f.

T

Tanks, 105
Taracouzio, T. A., 176
Taylor, G. E., 123
Temmerman, J., 114
Templer, Field-Marshal Sir Gerald, 67
Teske, Herman, 125
Thursby, Major P. D. F., 133, 135
Tientsin, 55, 59
Times (London), 37, 77, 134, 135, 137, 150, 156
Tito, 16, 20, 21, 24, 26, 27, 30, 45, 47, 48, 71, 105, 125, 126, 151, 161, 172, 175, 177, 178
Tongking, 62, 63, 154
Tsinan, 55, 58, 59
Tsingtao, 58
Tunisia, 23

V

Vaccarino, G., 22
Valeo, Francis, 57, 59
Viet minh, 23, 26, 34, 43, 47, 60 f., 66, 78 f., 118, 134, 138, 145, 146, 150
Viet minh Manual, 78 f., 145 f.
Vietnam Army, 63
Villa-Real, Lt-Colonel L. A., 155
Vinh, 64
Vitebsk, 65
Vollgraff, P. D., 22
Voroshilov, Marshal, 18

W

Warner, Denis, 43
Wehrmacht, *see* German Army
Willoughby, Major-General C. A., 17
Wingate, Major-General O., 113
Woodhouse, Colonel C. M., ix f., 34, 99, 115, 162, 165, 173, 174, 179, 180, 181

X

Ximenes, 43, 159, 160

Y

Yenan, 49, 53, 58
Yugoslavia, 15, 19, 20, 21, 24, 27, 30, 45, 46, 48, 52, 66, 71, 101, 105, 121, 123, 125, 143, 144, 172, 173, 174 f., 177, 178, 180, 181

Z

Zelenika, Colonel I., 123